# **Photo** Manual &
# **Dissection** Guide of

Fred **Bohensky**

SQUARE ONE'S ANATOMY SERIES

**Photo Manual and Dissection Guide of the Cat**
Fred Bohensky

**Photo Manual and Dissection Guide of the Fetal Pig**
Fred Bohensky

**Photo Manual and Dissection Guide of the Frog**
Fred Bohensky

**Photo Manual and Dissection Guide of the Rat**
Fred Bohensky

**Photo Manual and Dissection Guide of the Shark**
Fred Bohensky



ISBN 0-7570-0031-2 (Cat)



Printed in the United States of America

# CONTENTS

# DEDICATION

To my wife Esther, who gave up her summer vacation to type, proofread, and work along with me to see this book to its completion.

# PREFACE

The cat continues to be the favorite subject of dissection for courses in Comparative Anatomy, Anatomy and Physiology, and Mammalian Anatomy. In many instances, as in the pre-nursing curricula, human structures are taught without access to a human cadaver. Cats are the natural alternative. Their anatomy is similar to that of man, and comparisons to man will be made throughout.

The relatively large size of the cat makes dissection that much easier. It permits a study of muscles, blood vessels, and nerves which cannot be pursued with smaller, immature animals. Unlike the fetal pig, fully formed adult structures are observed.

The sheep heart, brain, and eye have been included to supplement similar structures of the cat and to facilitate comprehension. These are readily available and much larger than those of the cat. They enable one to make a more detailed as well as a comparative study.

The beginning Biology student is often overwhelmed when, at the start of the semester, he is handed a dissection specimen and instructed to follow the procedures outlined in the lab manual. He will encounter difficulty in relating the structures seen in the specimen before him with the diagrams in his lab manual. These are often poorly drawn, small, or inaccurately labeled. To the untrained eye, nerves, blood vessels, and smaller structures look very different from the figures as depicted. They are often idealized versions of "perfect" dissections.

"The Photo Manual and Dissection Guide of the Cat" presents full-page photographs revealing the various structures as they will appear to the student. Some are as big as the actual specimens, some are even larger. These have been carefully selected and labeled. Where doubt might exist as to the limits of an organ or a structure, these have been outlined on the photo with dotted lines, or their names have been printed directly on the organ. Some of the photos show the same areas in different views or in close-up. This will help students to pinpoint structures.

The cats used in these model dissections were young adults of average length. Arteries and veins were injected with red and blue latex dyes respectively. Generally, the legs, head, or tail have been included in the photograph to facilitate identification.

A Self-Quiz for students is found at the end of each unit. These consist of short answer type questions, definitions, and the labeling of review photographs. These worksheet pages may be removed from the book and submitted to the instructor for correction.

In this second edition several units have been added; others rewritten. The new units include an introductory section comparing the cat to man, an introduction to the nervous system, and a section on the ear. The unit on the skeleton has been greatly expanded. Throughout the book fully-labeled diagrams of human structures and systems are presented. These illustrations help to highlight the comparisons of the cat to the human.

The author wishes to acknowledge the help extended by the Media Production Center of The College of Staten Island. He is particularly grateful for the assistance of Mr. Joseph Rickard, the photographer, whose fine lenswork is seen throughout this manual.

Fred Bohensky
Staten Island, N.Y.
March, 1979

# INTRODUCTION: THE CAT AND MAN

For most students this is their first major dissection. A few words of introduction are in order.

Both the cat and man belong to the class of vertebrates known as *Mammals*. These are the most highly developed animal forms. While our major dissection subject is the cat, we shall be making constant reference to man in both the text and in the diagrams presented.

## THE MAMMALS

Since we are about to study the structures and functions of a type of mammal, we ought first to consider this class of vertebrates more closely.

Mammals range in size from minute shrews, which weigh only about two grams (0.002 kg) to giant blue whales, which weigh up to 115 tons (115,000 kg).

Besides mammals, the *Vertebrates,* or backboned animals, also include the:

— Fish
— Amphibians (frogs, toads, and salamanders)
— Reptiles (lizards, snakes, turtles, and crocodiles)
— Birds

The two chief mammalian characteristics which set these animals apart from the other classes of vertebrates are:

— Skin covered with hair or fur
— Milk producing glands (mammary glands) in the female to nurse the young

The most primitive living mammals belong to the order *Monotremata.* These animals, native to the Australian region, include the duck-billed platypus, and the spiny anteater. They lay a reptilian type of egg. When hatched, the young receive nourishment from the mother's mammary glands.

A more advanced form of mammal, belonging to the order of the pouched mammals, the *Marsupialia,* gives birth to live young *(viviparous).* However, the young are born at a very early stage of development and continue their maturation in a pouch, where they attach to nipples of the mammary glands. Australian kangaroos and American opossums are members of this order.

The true placental mammals known as the subclass *Eutheria,* to which almost all other mammals belong, includes the cat as well as man. Their embryos are retained in the uterus and are nourished by the placenta for a longer time and they emerge more highly developed. Most can walk and even run within a few days of birth; man is helpless for the longest period of time.

Man, *Homo sapiens,* belongs to the order *Primates* which also includes the apes and monkeys. The cat, whose generic and specific names are *Felis domestica,* belongs to the order *Carnivora.* The Carnivores include the lion, tiger, wolf, dog, and cat.

Cats are closely associated with civilized man as popular pets and valued for catching mice and rats. They were considered sacred by the ancient Egyptians, protected by law and even deified.

The average life span of the cat is about fifteen years. The *gestation* period, during which the embryo develops in the uterus, is approximately nine weeks (versus nine months for humans). There are two to five young per litter.

The cat is a comparatively large dissection specimen. Its muscles and internal organs are clearly visible. Its nerves and blood vessels are readily traced.

Comparisons to man will be made throughout. Most organs and tissues of the cat are structurally and functionally similar to those of man. Even their names are identical or quite similar. Whenever differences do exist, these will be pointed out. In addition, each major area studied will be illustrated with anatomical diagrams of similar human structures.

# ANATOMICAL TERMINOLOGY

Some basic biological terminology should be studied at this time. Familiarize yourself with the following words and learn to use them in referring to the location of the body parts of your specimen.

## DIRECTIONS OR POSITIONS

| | | |
|---|---|---|
| Anterior (Cranial) | - | toward the head |
| Posterior (Caudal) | - | toward the tail |
| Dorsal (Superior) | - | toward the backbone |
| Ventral (Inferior) | - | toward the belly |
| Lateral | - | toward the side |
| Medial | - | toward the midline |
| Proximal | - | lying near the point of reference |
| Distal | - | lying further from the point of reference |

Note: The terms in parentheses are synonymous only when referring to a quadruped such as a cat. In man these terms have different meanings (see diagrams at the end of this section).

## PLANES OR SECTIONS THROUGH THE BODY

| | | |
|---|---|---|
| Transverse (Cross Section) | - | perpendicular to the long axis of the body |
| Sagittal | - | a longitudinal section separating the body into right and left sides |
| Frontal (Coronal) | - | a longitudinal section dividing the specimen into dorsal and ventral parts |

DIRECTIONAL TERMS For the Cat and Man

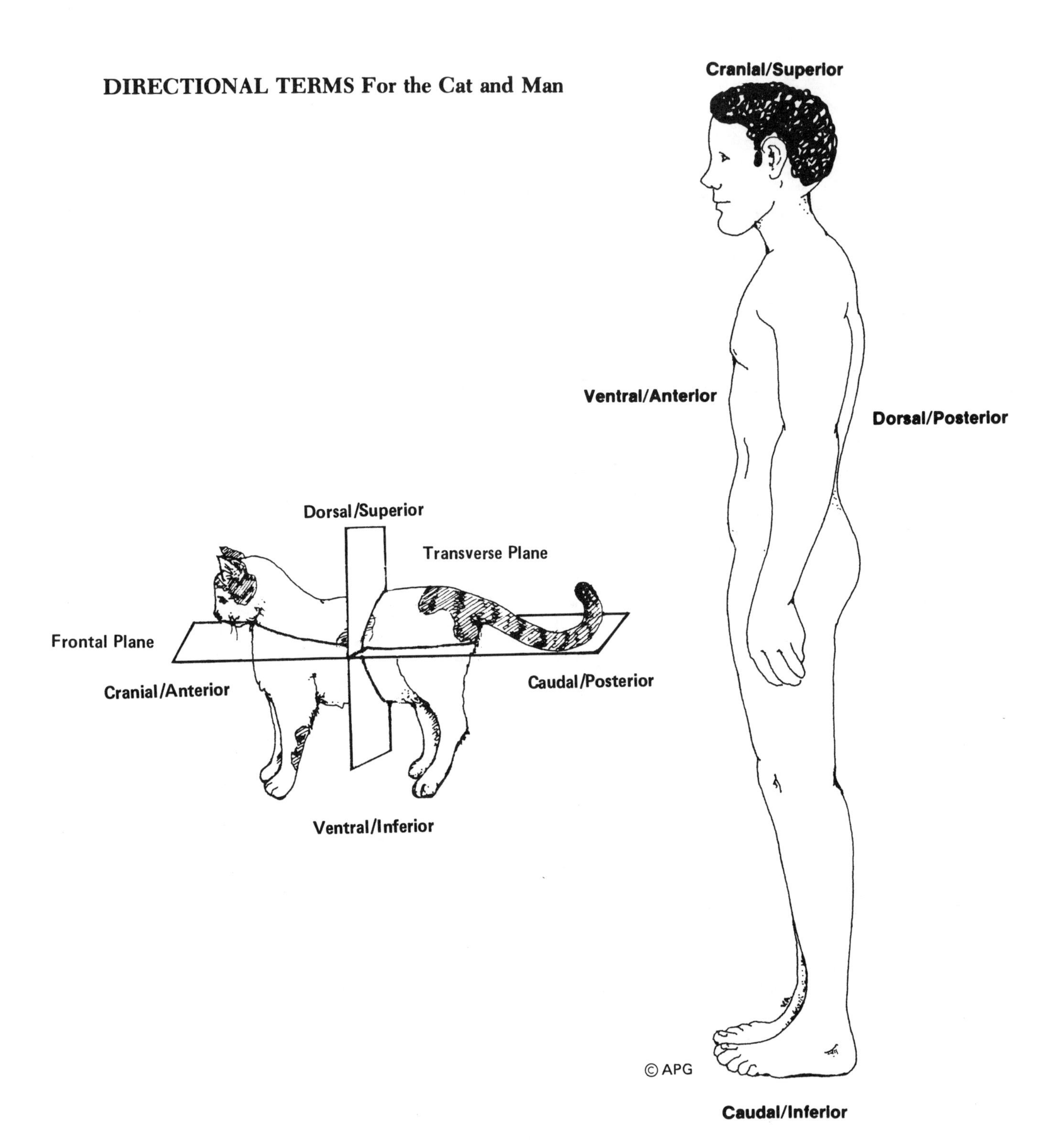

# EXTERNAL FEATURES

Examine the specimen you have received.

The cat, known scientifically as *Felis domestica,*belongs to the class *Mammalia.* All members of this group of vertebrates are unique in that they possess body hair or fur and the females develop mammary glands for nursing the young.

The student should now refer to the first photo.

The body is divided into four readily identifiable *areas:*

head (cranial)
neck (cervical)
trunk (thoracic and abdominal)
tail (caudal)

The paired *appendages* consist of:

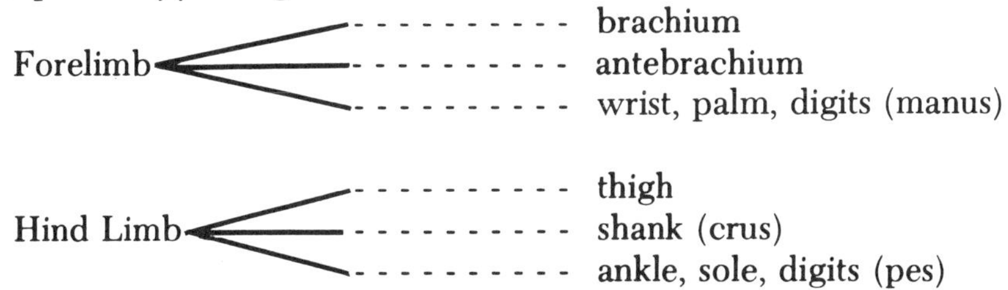


The last segment of each toe has a *claw.* This segment is hinged in such a way that the claw can be extended or retracted. The *tori* are epidermal thickenings on the walking surfaces of the feet. These friction pads form cushions upon which the cat walks. Seven tori are present on each forepaw and five on each hind paw.

On the head locate and identify the following:

**External Nares** - the nostrils
**Pinnae** - the large flexible flaps of tissue of the external ear.
**Nictitating Membrane** - spread the eyelids apart and observe this third eyelid in the medial corner of the eye. It can be extended laterally over the eyeball helping to moisten and cleanse the eye.
**Vibrissae** - long sensory hairs or tactile whiskers around the mouth

Note the *papillae of the mammary glands,* or teats, on the ventral surface of the trunk. They are arranged in two rows, usually hidden in the fur. There are about five pairs in the cat, but the number is subject to variation. Humans have a single pectoral pair. In pregnant and lactating females the mammary glands below the skin are well developed. Determine the number of papillae in your specimen. Are they found in both male and female cats?

Other areas to be recognized are the:

**Axilla** - between the thorax and arms
**Perineum** - the area around the urogenital and anal openings

The *anus,* the external opening of the rectum, is found ventral to the base of the tail. In males the *scrotum,* a pouch which contains the testes, is located anterior to the anus. In females, the *urogenital aperture* is located just ventral to the anus.

Is your specimen male or female? Examine a cat of the opposite sex and determine how its external features differ from those of your specimen.

**In the space below draw an outline of your cat and indicate all the external features you can find.**

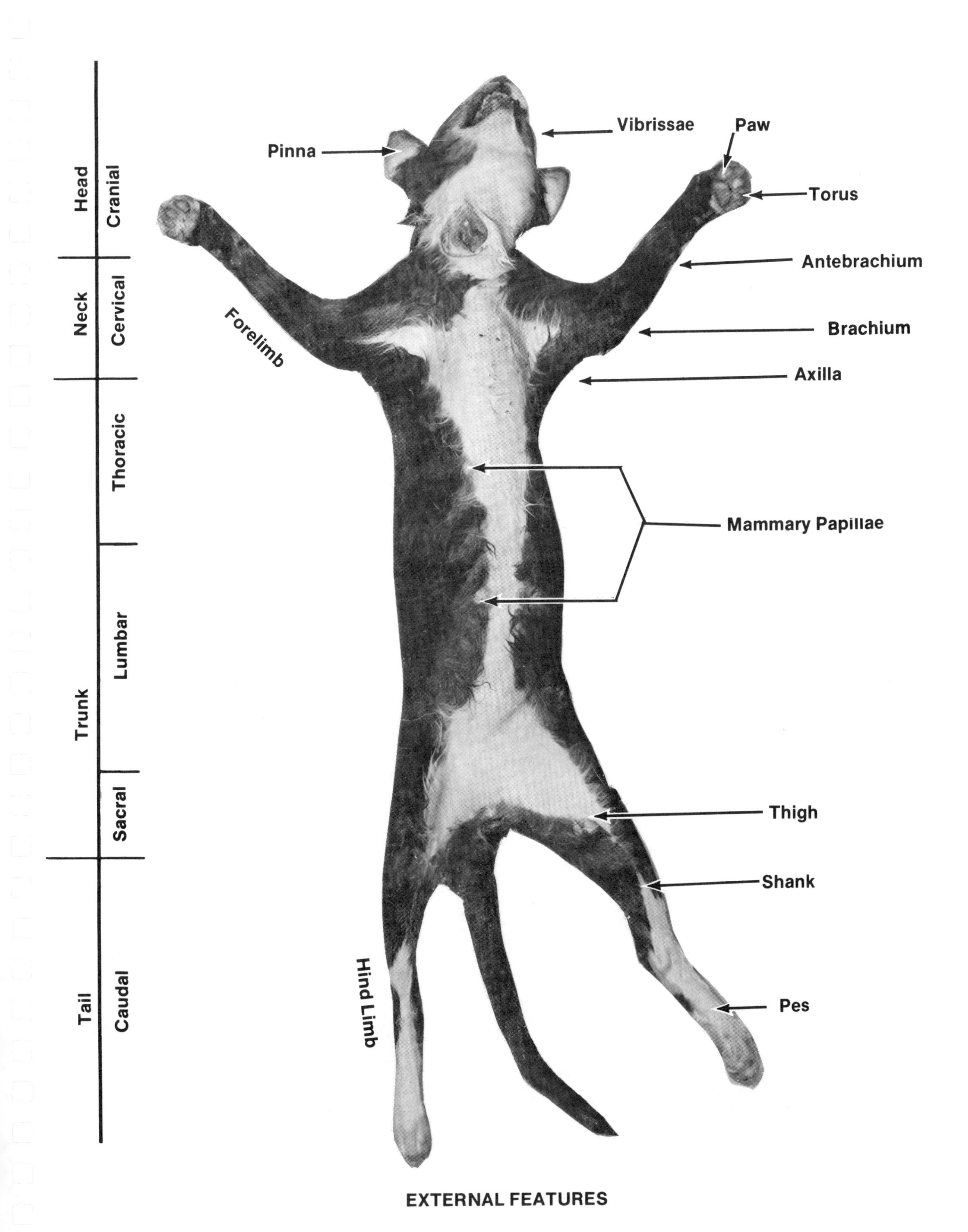


EXTERNAL FEATURES

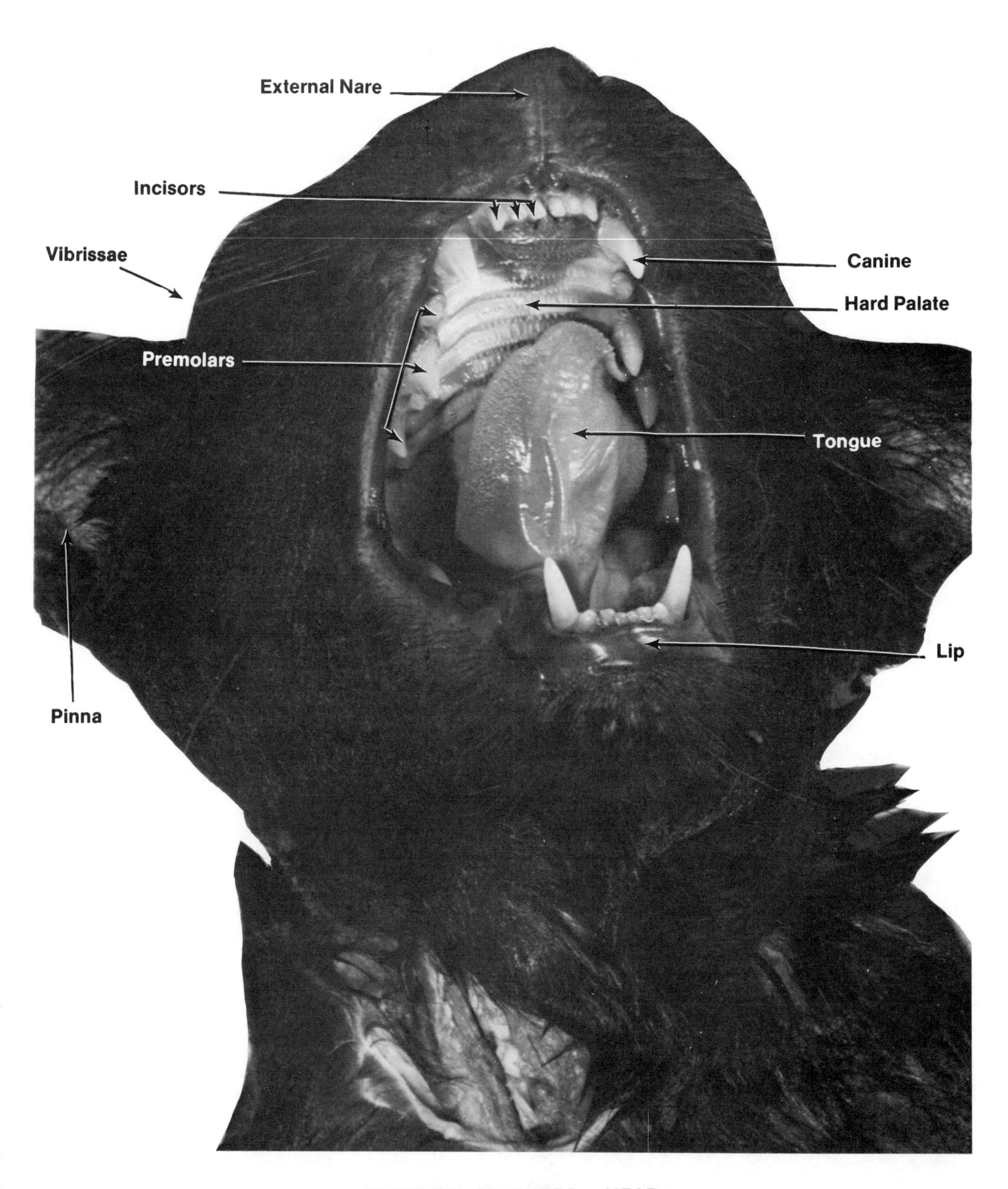


EXTERNAL FEATURES - HEAD

# THE SKELETON

The bony endoskeleton of the cat reveals many interesting features. A fully articulated and mounted cat skeleton should be available for student reference. These will appear as in the accompanying photo and diagram.

The human skeleton, fully mounted, should also be provided. A fully labeled diagram is included at the close of this unit. Throughout the study of the cat, compare its structures to those of man. You will note that the bones are named identically. Generally, it is only the number of bones that differ. Their arrangement in the various mammals is determined by differences in the method of support and locomotion.

The skeleton of all vertebrates is internal, known as the *endoskeleton.* It may be divided into two main areas.

**Axial skeleton** — includes the bones of the main longitudinal axis of the body.

**Appendicular skeleton** — the bones of the appendages plus their supporting girdles.

**Axial Skeleton**

The axial skeleton is composed of the:

- — Skull
- — Vertebral Column
- — Ribs
- — Sternum

**Skull** — The skull consists of the *cranium,* the bony vault protecting the brain, the *facial bones,* the *hyoid* bone, and the *auditory ossicles.*

**Cranium** — The human cranium is composed of 8 bones: 1 *frontal,* 2 *parietals,* 2 *temporals,* 1 *occipital,* 1 *sphenoid,* and 1 *ethmoid.* The cat's cranium consists of 11 bones.

| Cat | | Man |
|---|---|---|
| 2 | Frontal | 1 |
| 2 | Parietal | 2 |
| 1 | Interparietal | 0 |
| 2 | Temporal (Squamosal, Mastoid, Tympanic Bulla) | 2 |
| 1 | Occipital (Supraoccipital, Basioccipital, Exoccipital) | 1 |
| 1 | Basisphenoid } ...Sphenoid | 1 |
| 1 | Presphenoid } | |
| 1 | Ethmoid | 1 |
| 11 | | 8 |

**Facial Bones** — The cat has 15 facial bones, man has only 14. The *premaxilla* bones, between the maxilla and nasal bones, are not found in man. Identify and learn the names of the facial bones: *maxilla, zygomatic, lacrimal, nasal, vomer, palatine,* and *mandible.*

| Cat | | Man |
|---|---|---|
| 2 | Nasal | 2 |
| 2 | Premaxilla | 0 |
| 2 | Maxilla | 2 |
| 2 | Lacrimal | 2 |
| 2 | Malar (Zygomatic) | 2 |
| 1 | Vomer | 1 |
| 2 | Palatine | 2 |
| 2 | Mandible | 1 |
| 0 | Inferior Conchae | 2 |
| 15 | | 14 |

**Hyoid** — This bone in man does not articulate with any other bone. Its superior aspects support the musculature of the tongue, posteriorly it supports the cartilage of the larynx. It is suspended by ligaments from the styloid processes of the temporal bones.

In the cat, the hyoid is more extended and consists of *eleven bones.* The *anterior horn* is attached to the tympanic bulla of the temporal bone, the *posterior horn* is attached to the thyroid cartilage of the larynx.

**Auditory Ossicles** — Three pairs

The bones of the middle ear are embedded in the tympanic bullae of the *temporal* bones.

They are the — *malleus* (hammer)
— *incus* (anvil)
— *Stapes* (stirrup)

**Vertebral Column** — The outstanding characteristic of the vertebrates is the possession of a backbone or *vertebral column.* It serves as an attachment for the muscles of the back and its support. The soft delicate *spinal cord* runs through the vertebral bones, and is protected by them. The vertebral column of the cat is composed of 49-53 bones, that of man only 26.

A typical vertebral bone consists of the *body* or *centrum, neural arch* with *spine,* a pair of lateral and transverse *processes,* and the posterior *articular facets* at the point where the vertebrae meet. A pad of cartilage, the *intervertebral disc,* forms a protective cushion between adjacent vertebrae.

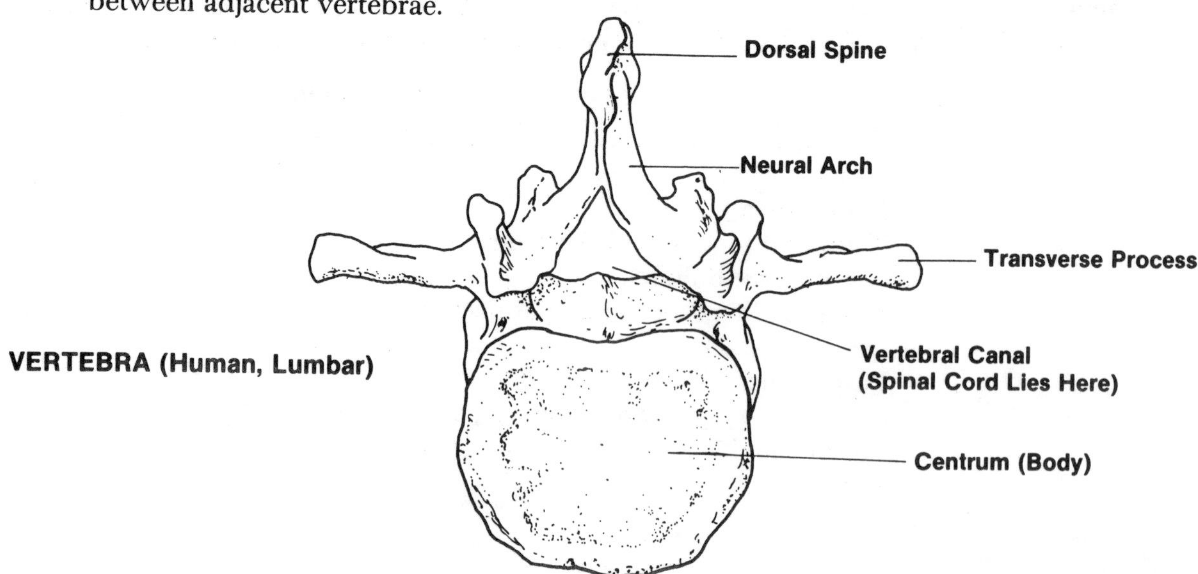


VERTEBRA (Human, Lumbar)

Five sections of the vertebral column are readily identified:

1. **Cervical** — These 7 bones are in the neck region. The topmost two are the *atlas* and the *axis*. They permit free movement and rotation of the head. Virtually all vertebrates, even tall giraffes, have 7 cervical vertebrae.
2. **Thoracic** — They number 12 in man, 13 in the cat. They have prominent dorsal spines and *articular costal* facets for the attachment of ribs.
3. **Lumbar** — These number 5 in man and 7 in the cat. These are characterized by massive centra, reduced dorsal spines, and long transverse processes.
4. **Sacral** — In man 5 sacral vertebrae are fused into one bone, the *sacrum*. The cat has only 3.
5. **Caudal** — There are 21-25 caudal vertebrae in the cat extending out into the tail. In man 3-5 bones are fused to form the *coccyx*.

Variations in the number of vertebrae:

| Region | Cat | Pig | Man |
|---|---|---|---|
| Cervical | 7 | 7 | 7 |
| Thoracic | 13 | 14-15 | 12 |
| Lumbar | 7 | 6-7 | 5 |
| Sacral (Fused) | 3 | 4 | 5 |
| Caudal | 21-25 | 20-23 | 3-5 (Coccyx) (Fused) |

**Ribs** — In man there are 12 pairs of ribs. Of these, the upper 7 pairs are known as *"true" ribs*. They articulate with the thoracic vertebrae as well as with the *Sternum*, or breasbone. The attachment to the sternum is by way of *costal cartilage*, not bone. The next 3 are called *"false" ribs*. They do not articulate with the sternum directly only by way of the seventh rib. The last two pairs of ribs are called *"floating" ribs*. They articulate with the thoracic vertebrae, but do not reach the sternum at all. There are 13 pairs of ribs in the cat.

**Sternum** — The sternum of the cat is elongated and consists of eight separate units called *sternebrae*. The first sternebrae is the *manubrium*, the next six constitute the *body* of the sternum, and the last sternebra is the *xiphisternum* (or *xiphoid process*). A small piece of cartilage, the *xiphoid cartilage*, extends posteriorly.

The sternum of man is must reduced and consists of three fused portions, the superior *manubrium*, the central *body* (or *gladiolus*), and the inferior *xiphoid process*.

The *hyoid* bone and the three pairs of *auditory ossicles* are generally considered as parts of the axial skeleton, although, embryologically they have quite different origins and are traced to the gill apparatus, jaw, and pharynx of fish.

**Appendicular Skeleton**

The appendicular skeleton consists of the *pectoral girdle* and the attached *forelimbs*, and the *pelvic girdle* and the attached *hind limbs*.

**Pectoral Girdle** — In man there are two pairs of bones which comprise this structure, the two *scapulas* and two *clavicles*.

The cat's clavicles are slender and curved. They do not articulate with any portion of the skeleton. They lie embedded in the shoulder muscle.

**Forelimbs** — In man each forelimb is composed of the *humerus* (upper arm), *ulna* and *radius* (lower arm), 8 *carpal* bones (wrist), 5 *metacarpals* (palm), and 5 *digits* composed of 14 *phalanges*.

The bones of the cat's forelimbs are homologous to those of man and are named identically. The number of carpal bones, however, in the cat is only seven, not eight as in man.

**Pelvic Girdle** – The pelvic girdle of the cat as of man is formed by paired bones the *innominate*, or *os coxa.* Each is, in turn, three separate bones which fused during fetal development. They are the *ilium* laterally, the *ischium* posteriorly, and the *pubis* ventrally. A pad of cartilage, the *pubic symphysis*, lies at the mid-ventral juncture of the pubic bones and unites them ventrally. Posteriorly, the sacrum articulates with the two ilium bones to complete the entire pelvic girdle. The articulation is known as the *sacroiliac* joint. A spherical depression, the *acetabulum* serves as the point of articulation for the hind limb. It is the "socket" that receives the ball-shaped head of the femur.

The pelvic girdle is longer and narrower in the cat than in man. This is an adaptation to its method of locomotion. Man the *biped*, walking on two limbs, requires an enlarged pelvis to support the entire axial skeleton and the upper appendages. The cat, a quadreped, is able to distribute its weight more evenly and achieves a better balance upon its four legs. Thus, less strain is placed upon the smaller pelvis.

**Hind Limbs** – In man each leg is composed of the *femur* (thigh bone), *patella* (knee cap), *tibia* and *fibula* (skin and calf bones), 7 *tarsal* bones (heel), 5 *metatarsals* (sole of foot), and 5 *digits* composed of 14 *phalanges*.

In the cat the bones are similar to those of man. The primary difference is in the number of digits. While the cat has five digits on each foreleg, it has only four on each hind leg.

**Feet**

Observe the position of the paws in walking. In the cat the gait is called *digitigrade*, which means that the animal walks on its digits with the remainder of the foot elevated. Humans walk *plantigrade*, on the entire sole of the foot. Horses and cattle walk on hooves and are therefore *unguligrade* walkers.

**Teeth**

An animal's diet is revealed by its *dentition* pattern. This refers to the types of teeth the animal possesses, their number, and arrangement.

In the cat, sharp and pointy *incisors* and *canines* dominate the front of the mouth, and *premolars* at the rear. A single *molar*, reduced in size is located most posteriorly in each quarter of the jaw. The accompanying photo close-up of the open-mouthed cat reveals its dental pattern. These are characteristic of a *carnivorous* diet. Horses and cows possess incisors for shearing grass and other vegetable matter and rows of large flattened molars with broad grinding surfaces toward the rear of the mouth. This is typical of the *herbivorous* pattern. The human dentition indicates an *omnivorous* diet. It combines sharp pointy incisors and canines with grinding premolars and molars.

The dental formulas of the adult cat and man are compared below:

**Cat** – $I\frac{3}{3},\ C\frac{1}{1},\ P\frac{3}{2},\ M\frac{1}{1}$

**Man** – $I\frac{2}{2},\ C\frac{1}{1},\ P\frac{2}{2},\ M\frac{3}{3}$

The letters refer to the types of teeth:
I – Incisor, C – Canine, P – Premolar, M – Molar

The upper set of numbers refers to the number of teeth in half of the upper jaw, the lower set of numbers refers to the number of teeth in half of the lower jaw. Thus the total of teeth in adult cats is 30 and 32 in man.

In summary, some of the major differences between the skeletal systems of the cat and man are:

1. Variations in the number of *vertebrae.*
2. Man has 22 separate *skull* bones, whereas the cat has 35-40.
3. There are 13 pairs of *ribs* in the cat, not 12 as in man.
4. The *clavicles* of the cat are reduced. They do not articulate with the *sternum* or *scapula* as in man, but are embedded in muscle.
5. The cat has 7 *carpal* bones, man has 8.
6. The cat has 5 digits on the foreleg, but only 4 on the hind leg.

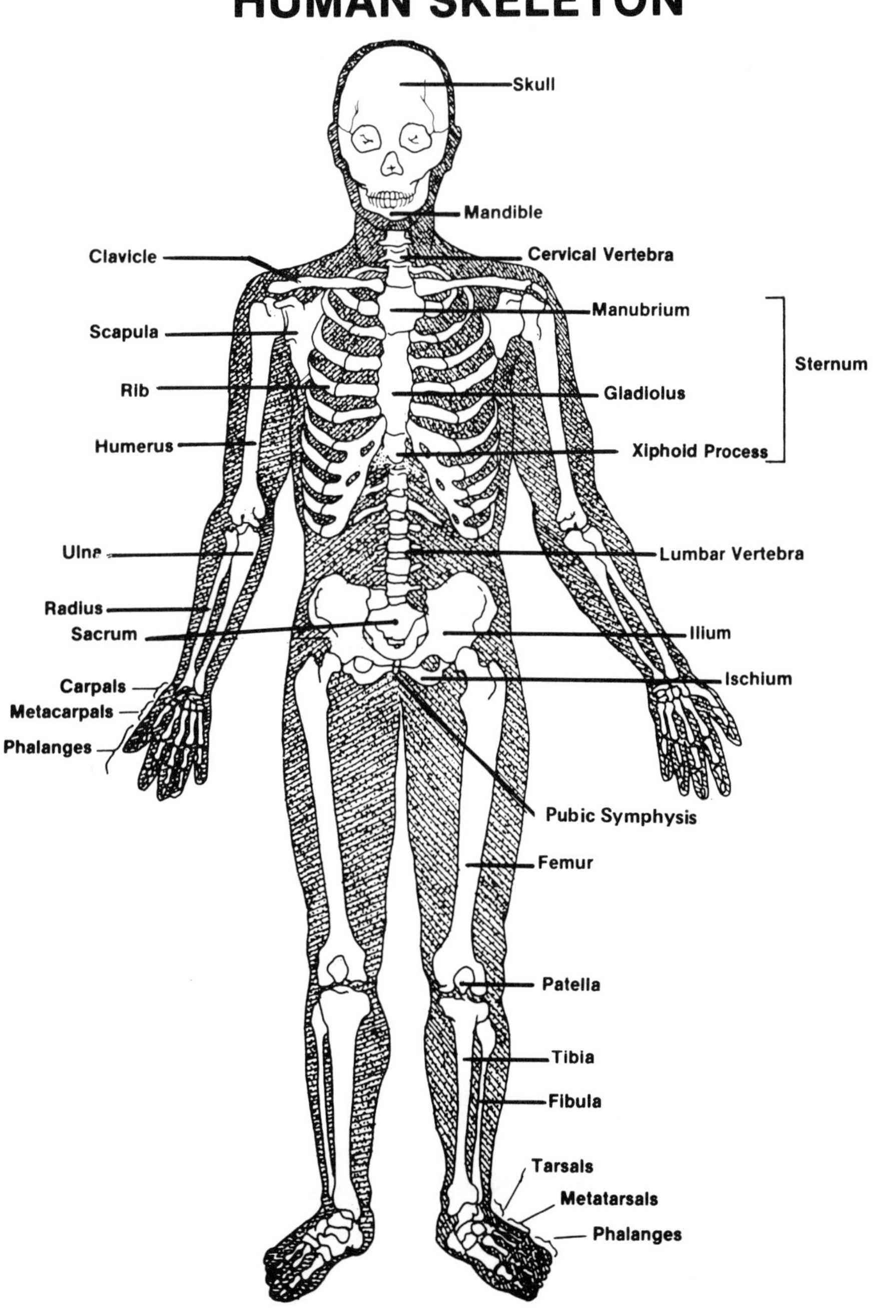

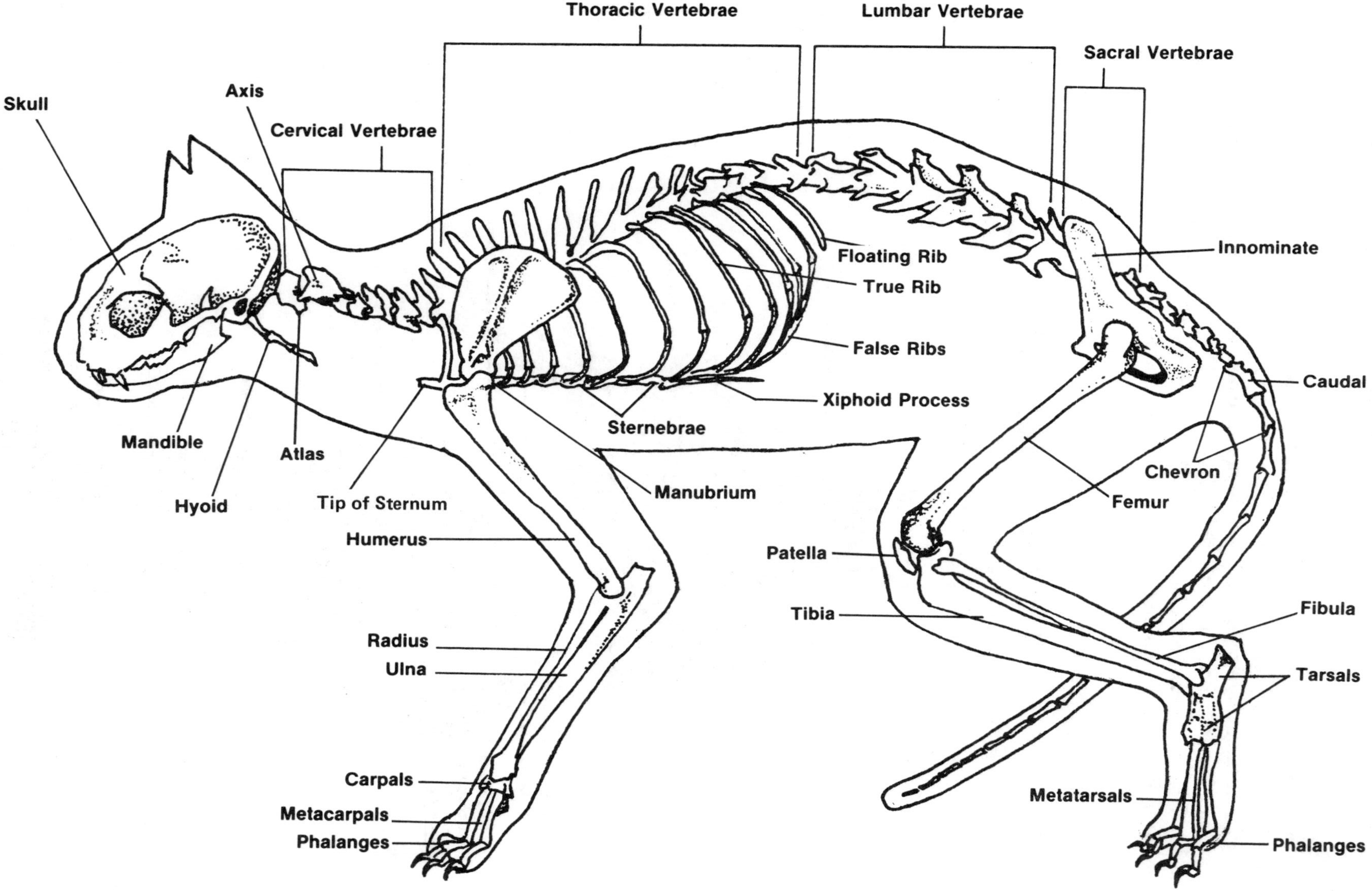

Thoracic Vertebrae
Lumbar Vertebrae
Sacral Vertebrae
Skull
Axis
Cervical Vertebrae
Innominate
Floating Rib
True Rib
False Ribs
Caudal
Xiphoid Process
Sternebrae
Mandible
Atlas
Chevron
Hyoid
Tip of Sternum
Manubrium
Femur
Humerus
Patella
Tibia
Fibula
Radius
Ulna
Tarsals
Carpals
Metatarsals
Metacarpals
Phalanges
Phalanges

# CAT SKELETON

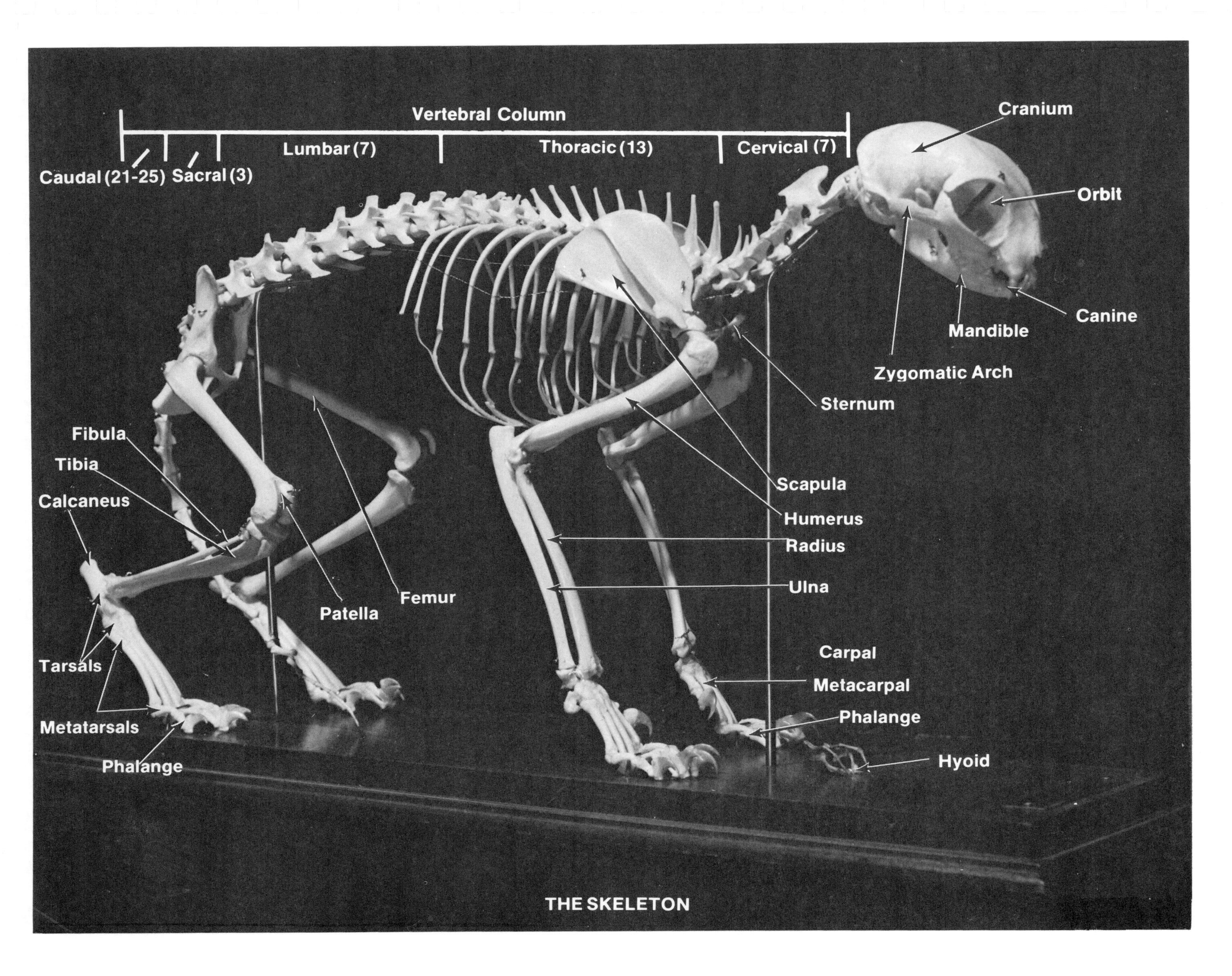


THE SKELETON

# GENERAL DISSECTION HINTS

The term "dissection" means more than merely cutting your specimen apart. It is a refined method of seeking, exposing, identifying, and studying the internal anatomy. It helps to bring into view structures not readily seen.

Use your *scalpel* sparingly. Rely primarily upon your dissecting needles. They are especially helpful in tracing blood vessels and nerves and in separating muscles. As organs are exposed, study the associated blood vessels and nerves. Leave them intact unless directed to do otherwise. Check often with the photos in the manual. Confirm the names and location of body parts of your specimen.

When using your *scissors*, advance with the rounded, blunt end, not the sharp, pointed end. Your *forceps* should be strong, able to hold on to thick muscle, yet fine enough to grasp narrow nerves. It is advisable to have more than one type of forceps. Move organs aside with your fingers or with a blunt probe.

In order to protect yourself against the odor and the effects of the preservative solution upon your hands it is suggested that you apply lanolin or vaseline at the outset or wear thin rubber gloves. Line your dissection pan with paper towels in order to absorb excess fluids, as a storage for structures removed, and to facilitate cleaning up at the close of the session.

At the end of each class, wrap the cat in wet paper towels before returning it to the plastic bag. Twist the top of the bag and close tightly with a rubber band. These procedures will protect your specimen from drying out between dissection sessions. Remove the paper towels lining the dissecting pan, together with any structures removed, and discard.

In order to further preserve the softness and texture of cat muscles and organs, apply the following solution with a one inch paint brush at the close of each session.

| | |
|---|---|
| Carbolic Acid (Phenol) crystals | 30 grams |
| Glycerin | 250 ml. |
| Water | 1000 ml. |

# MUSCULAR SYSTEM, AN OVERVIEW

Skeletal muscles enable the body to move. Most are firmly anchored to bone at one end, the *origin* of the muscle, while the other end is attached to the bone to be moved, and is known as the *insertion*. The fleshy central portion is termed the *belly*. The ends of a muscle are attached to bone most often by means of a narrow band of connective tissue called a *tendon*. They may also be joined directly to the *periosteum* of the bone. Finally muscles may be united with each other or to a bone by means of a broad, flat sheet of tendinous tissue known as an *aponeurosis*.

As you dissect, locate the origins and insertions of the muscles studied. Then free the muscle from other muscles and from the nerves and blood vessels associated with it. The fine, transparent connective tissue which binds adjacent muscles is *deep fascia*, while tougher and more fibrous *superficial fascia* connects the skin to the muscles below. When the muscle has been freed, pull it gently. This will duplicate the muscle's normal contraction. Observe which bones or organs are moved, and which are relatively stable.

The *action* of a muscle results from its contraction. Muscles are usually arranged in *antagonistic* pairs. This means that while a muscle will cause a structure to move in one direction, the antagonist will cause it to move in the opposite direction.

**ACTIONS OF MUSCLES:**

**Flexion** - to bend at a joint decreasing the angle at that joint; examples: elbow or knee joint
**Extension** - to straighten joint increasing the angle at that joint
**Adduction** - to move appendage toward sagittal midline; example: lowering arms from shoulder level to rest at sides
**Abduction** - to move appendage away from sagittal midline; example: raising arms from rest at sides to shoulder level
**Supination** - to turn palm of hand upward
**Pronation** - to turn palm of hand downward
**Rotation** - to move a structure about a point; example: turning head from side to side
**Circumduction** - when the distal end of a limb describes a circle while the proximal end remains fixed, as the vertex of a cone; example: the movement of the extended arm in drawing a circle on the blackboard

# BEGINNING THE DISSECTION

Generally, your cat specimen will be preserved with its limbs extended. This, together with its relatively large size, enables you to proceed without tying the animal to the dissection tray.

Turn the animal so that its ventral surface is upward. Use your scalpel, forceps, and blunt probe. Proceed as in the photo.

1. Make a mid-ventral incision from the jaw to the external genitalia. Cut through skin only, leave the underlying muscles intact.

2. Lift the skin with forceps and separate the skin from the underlying muscles. You will note that the two are held together by a white fibrous connective tissue known as the *superficial fascia.* Cut the fascia as you loosen the skin. In female spécimens make a circular incision around each of the papillae of the mammary glands.

3. Continue to reflect the skin toward the dorsal side. Use the back of your scalpel, a blunt probe, or your fingers to facilitate separation.

4. Cut along the medial surface of the forelimbs and hind limbs and extend to the paws.

5. Leave the skin intact around the urogenital and anal areas *(perineum).* Remove the skin from the proximal 1/3 of the tail.

6. Turn the cat over. Complete the skinning of the limbs and the entire dorsal surface from the base of the skull, neck, thorax, and abdomen, to the proximal 1/3 of the tail. Do not discard the skin. Use it to wrap the cat, in addition to wet paper towels, at the close of each dissection session. At this point the only areas still covered by skin are the: head, mammary papillae, paws, perineum, and the distal portion of the tail.

Your cat should now appear as the one in the photo on page 37.

Note: If your specimen is a pregnant or lactating female, the *mammary glands* will appear as a pair large, longitudinal, glandular masses along the ventral sides of the abdomen and thorax below skin. They should be removed with the skin.

Look for light brown fibers adhering tightly to the underside of the skin. These are the *cutaneous muscles.* They include the:

**Cutaneous Maximus** - This muscle covers most of the sides of the body in the thoracic and abdominal areas. It serves to twitch the skin to avoid irritants. It originates from muscles in the axilla, the thorax, and abdomen, and inserts on the skin. It is not found in man.

**Platysma** - This is another cutaneous muscle. It is found on the lateral surface of the head and neck. It moves the skin on the neck and the face. It originates from the mid-dorsal area over the neck and inserts in the skin of the face near the ears, eyes, and mouth.

As you continue the dissection trim the fascia, other connective tissue and fat covering the muscles. Observe the direction in which the muscle fibers lie. The fibers of a single muscle are generally oriented in only one direction. Look for natural separations between muscles, then slit the fascia between muscles with a dissecting needle or a scalpel. Caution: Do not cut the muscle fibers, only the connective tissue between them.

When the study of superficial muscles has been completed, *transect* the muscles. This is done by cutting them at right angles to the directions of the fibers at the belly area and folding them back, or *reflecting* them, to their origins and insertions. This will reveal the deeper muscle layers.

We shall begin by examining the superficial muscles on the ventral surface, then on the dorsal surface. Each will be followed by a review. The deeper muscles will then be studied. A Self-Quiz for students will be found at the end of the unit on the Muscular System.

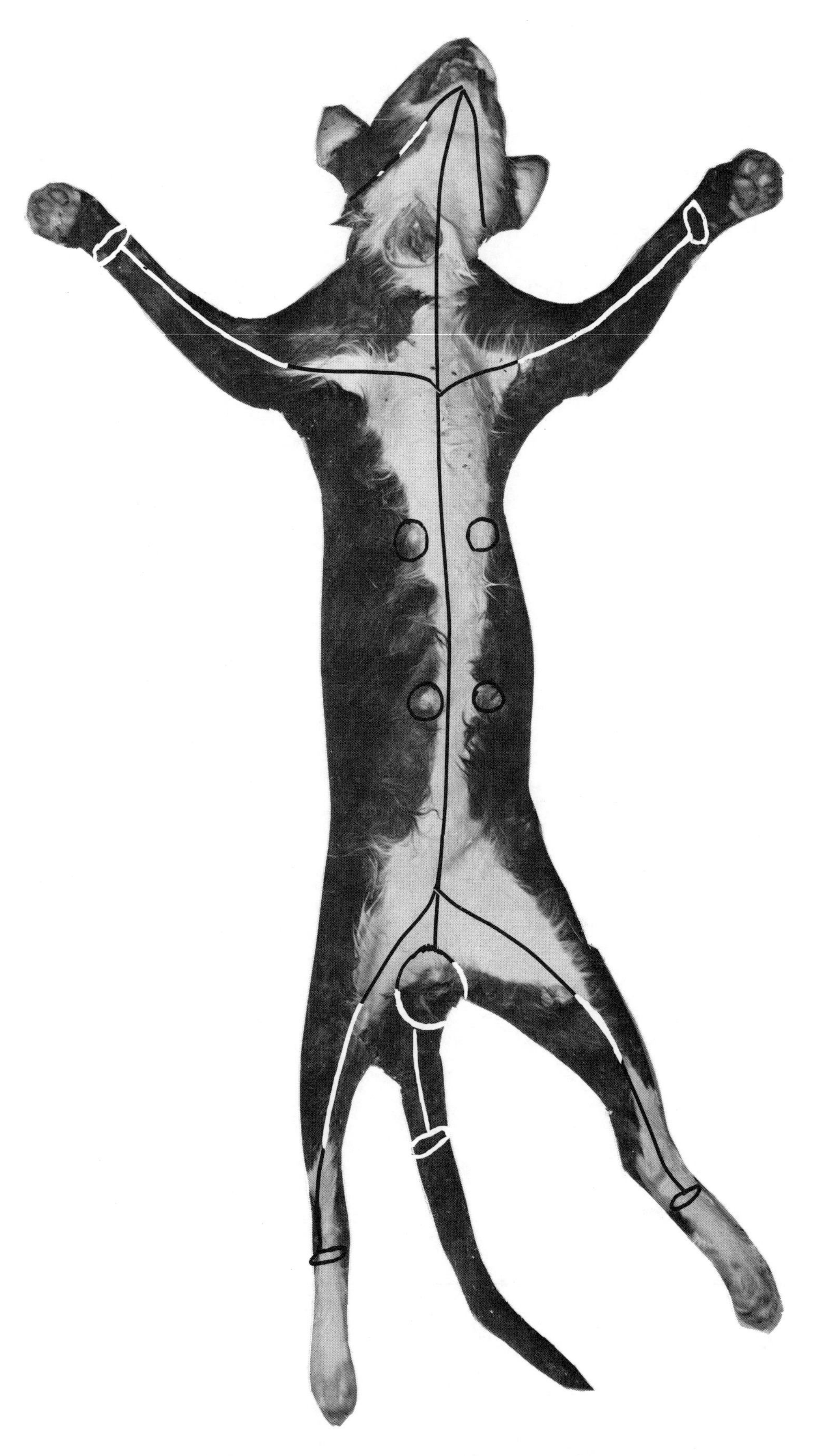

BEGINNING THE DISSECTION - REMOVAL OF SKIN

# SUPERFICIAL MUSCLES - THORAX, VENTRAL VIEW

Turn the cat to expose the ventral surface of the thorax and shoulder as in the photo.
Note: A view of the entire superficial ventral musculature is seen in the photo on page 37.

The large triangular mass of chest muscles comprising the *pectoral* group dominates this area. It originates in the sternum and inserts upon the humerus. It acts to pull the humerus toward the chest (adduction) and caudally. In man there are only two pectoral muscles, while in the cat there are four.

**Pectoantebrachialis** - This is the most anterior of the four. It lies as a narrow band across the top of the chest. It originates in the anterior portion of the sternum and inserts on the fascia of the ulna near the elbow. Clear the fascia connecting this muscle to its neighbors. The fibers of the pectoantebrachialis are readily identified since they run at right angles to the sternum, while the remaining pectoral muscles meet the sternum obliquely. This muscle is not found in man.

**Pectoralis Major** - Its anterior portion lies beneath the pectoantebrachialis, while its larger portion lies posterior to it. It too arises at the anterior sternum and inserts upon the humerus. In humans it is the largest of the pectoral muscles; thus its name.

**Pectoralis Minor** - This is the largest of the pectoral muscles of the cat. It lies posterior to the pectoralis major. In humans it is much reduced and lies deep to the pectoralis major. It originates from the mid and posterior sternum and inserts on the humerus.

**Xiphihumeralis** - This is the most posterior of the pectoral muscles. It originates on the most posterior portion of the sternum, the xiphoid process, and inserts on the humerus together with the pectoralis major and the *latissimus dorsi,* a muscle of the back. There is no muscle corresponding to the xiphihumeralis in man.

The most anterior muscle of the shoulder and arm seen in the photo is the *clavobrachialis*. It will be described on page 29.

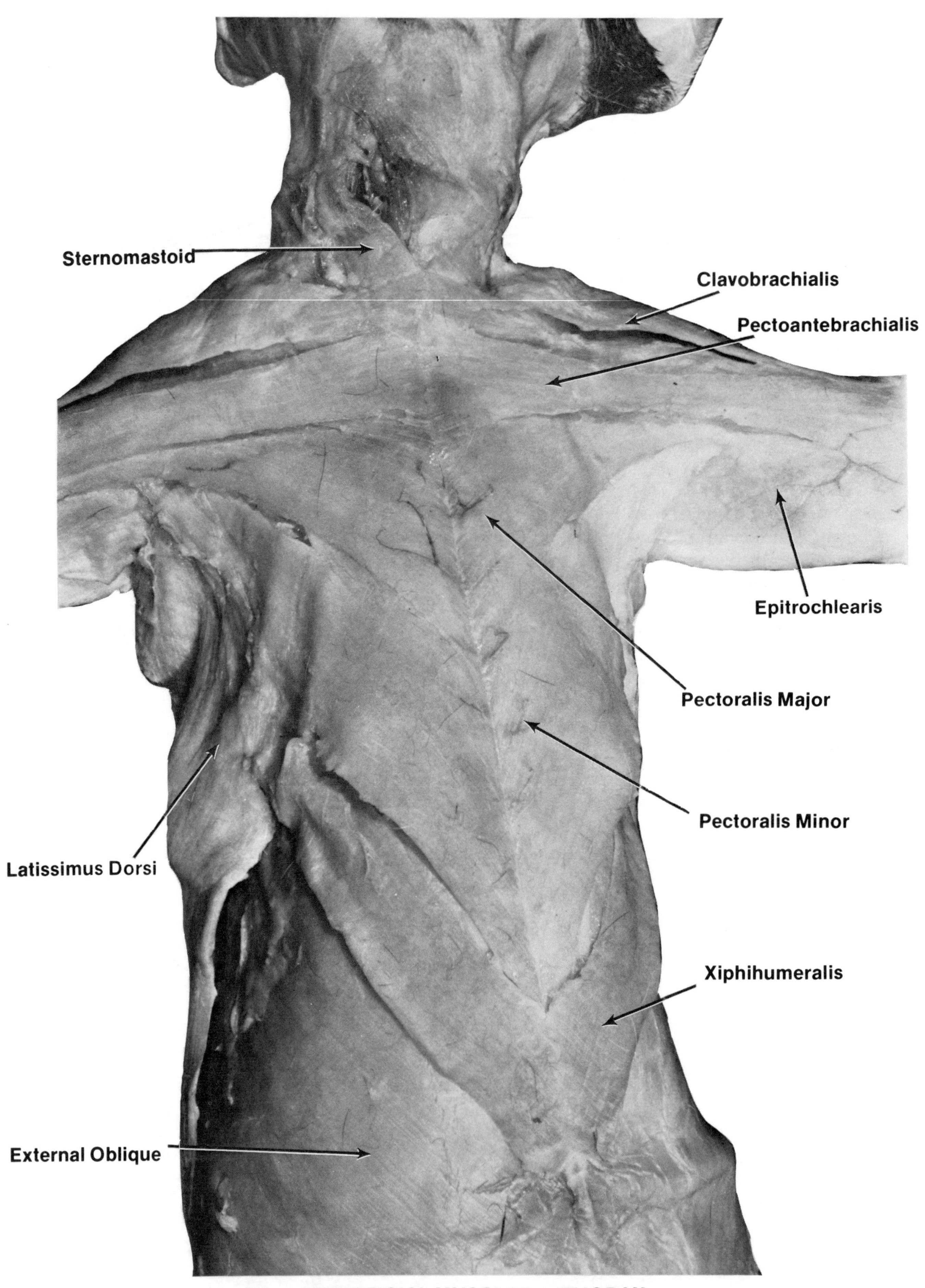


SUPERFICIAL MUSCLES - THORAX

# SUPERFICIAL MUSCLES - ABDOMEN, VENTRAL VIEW

The abdominal area is not protected by a bony structure as is the thorax. The abdominal organs are held in place by the pressure of the lateral abdominal muscles. They compress the abdominal wall and aid in flexion of the trunk. The abdominal muscles include the:

**External Oblique** - This is a thin broad sheet of muscle covering the ventral and lateral abdominal surfaces. It is the outermost of the three lateral abdominal layers. It originates on the posterior ribs and the *lumbodorsal fascia,* an aponeurosis on the dorsal surface, and inserts on an aponeurosis along the mid-ventral surface. The *linea alba,* a white line of conective tissue along the mid-ventral surface represents the fusion of the aponeuroses of the right and left sides. The fibers of the external oblique extend caudally and ventrally in an oblique direction across the abdominal surface.

**Internal Oblique** - Lift the edge of the external oblique where it joins the aponeurosis as in the photo, and expose the second layer of abdominal muscles, the internal oblique. Its fibers run in a direction opposite to those of the upper layer, namely, ventrally and anteriorly.

**Transversus Abdominis** - This is the innermost of the abdominal muscle layers. Its fibers extend ventrally and slightly caudally, almost parallel to those of the external oblique. It arises from the lower ribs and the lumbar vertebrae and inserts along the linea alba by an aponeurosis. The arrangement of the fibers of the three layers gives the abdominal wall its strength. Below the transversus abdominis lies the thin glistening membrane, the *parietal peritoneum* which lines the abdominal cavity.

**Rectus Abdominis** - In the mid-ventral area, on either side of the linea alba, lie two parallel muscles. They extend from the pubis cranially to insert on the upper ribs and sternum. For much of their course they lie between the aponeurosis of the internal oblique and the transversus abdominis.

**Latissimus Dorsi** - Although, as the name indicates, this is primarily a muscle of the dorsal surface, it is very prominent on the lateral ventral surface as well. It is readily seen in several photos. It is a flat, broad muscle, triangular in shape with an extensive origin. It arises from aponeuroses along the mid-dorsal line of the posterior thoracic region and from most the lumbar region. It covers the lateral surface of the body in this area. It extends ventrally to insert on the humerus. It gives to the humerus great power for pulling backward when the cat is running or climbing.

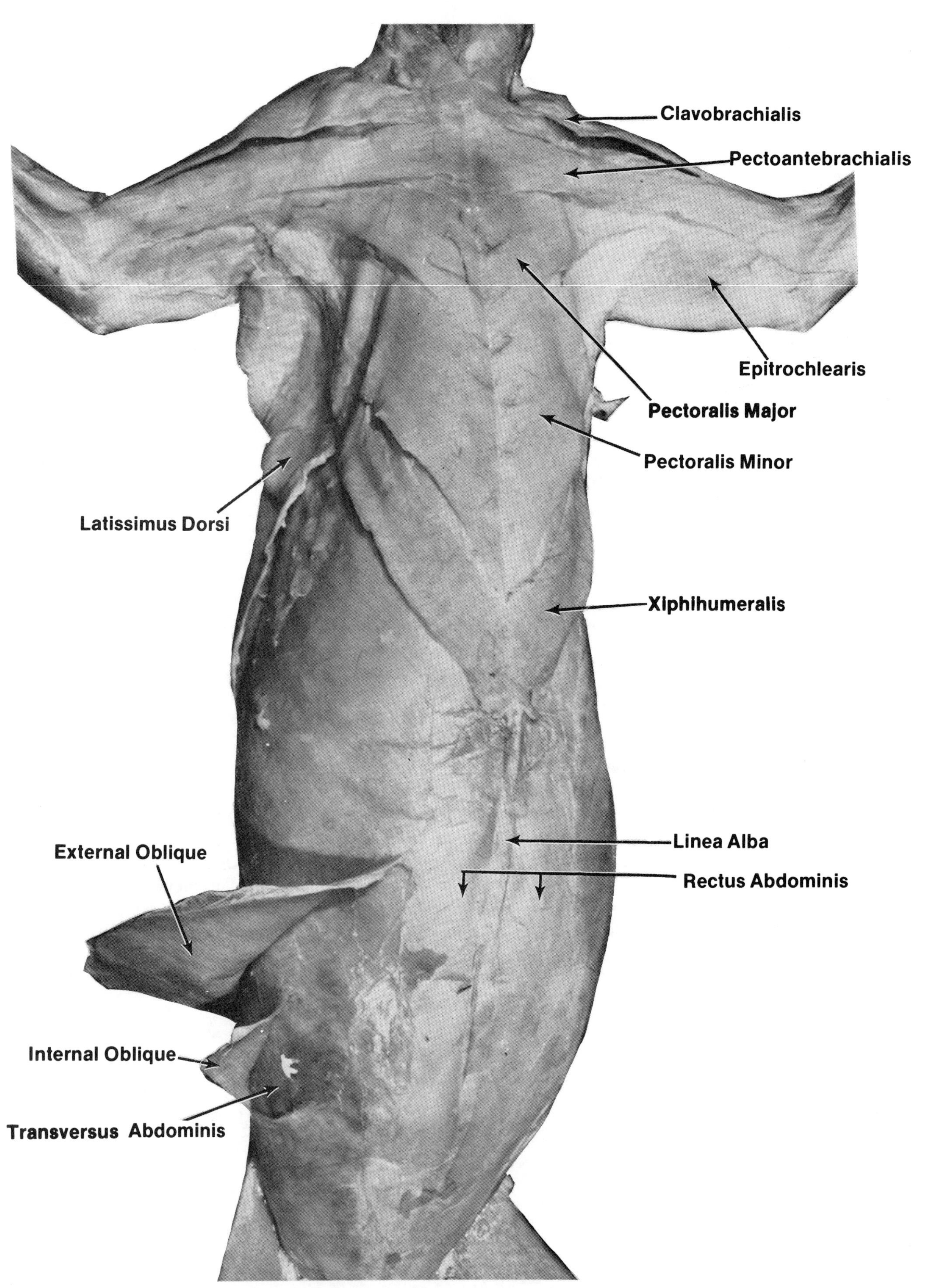


SUPERFICIAL MUSCLES - THORAX AND ABDOMEN

# SUPERFICIAL MUSCLES
# HEAD AND NECK, VENTRAL VIEW

Carefully remove the skin from the ventral and lateral surfaces of the head and neck as in the photo. This exposes prominent blood vessels, salivary glands, as well as facial muscles.

**Clavobrachialis** - This is the most anterior of the shoulder and arm muscles. It will be described along with the muscles of the forelimb on page 29.
**Sternomastoid** - This is the large "V" shaped band of muscle on the ventral and lateral surfaces of the neck.
**Cleidomastoid** - This muscle lies lateral to the sternomastoid. In man they are combined as a single muscle, the *sternocleidomastoideus*. These muscles enable the cat to flex its head and to bend it to the right or left.

The names of these muscles indicate their origins and insertions. The sternomastoid originates on the namubrium of the sternum and inserts upon the mastoid process of the temporal bone. The cleidomastoid originates from the clavicle and similarly inserts upon the mastoid process.

**Sternohyoid** - Two bands of muscle tissue run along the mid-ventral line in a posterior to anterior direction. They originate in the sternum and insert upon the hyoid cartilage of the neck.

In the photo, you can see the *trachea* and its cartilage rings in the space between the sternohyoid muscles.

**Sternothyroid** - This muscle lies deep to the sternohyoid and lateral to it.
**Digastric** - Below the inner edge of the mandible one can see the "V" shaped digastric muscle. It extends from the occipital and temporal bones to the mandible and acts to lower the jaw.
**Mylohyoid** - The fibers of this muscle run transversely between the digastric muscles in the mid-ventral area. The ends of the muscle pass deep to the digastric. It acts to raise the floor of the mouth.

The *masseter* and *temporalis* muscles will be described on the next page when the lateral view of the head is discussed.
The blood vessels seen in the photo are the:

**External Jugular Vein**
**Anterior Facial Vein**
**Posterior Facial Vein**
**Transverse Jugular Vein**

The *parotid and submaxillary salivary glands* are seen. They too are much more prominent in the lateral view.
Several *lymph glands* are also present.

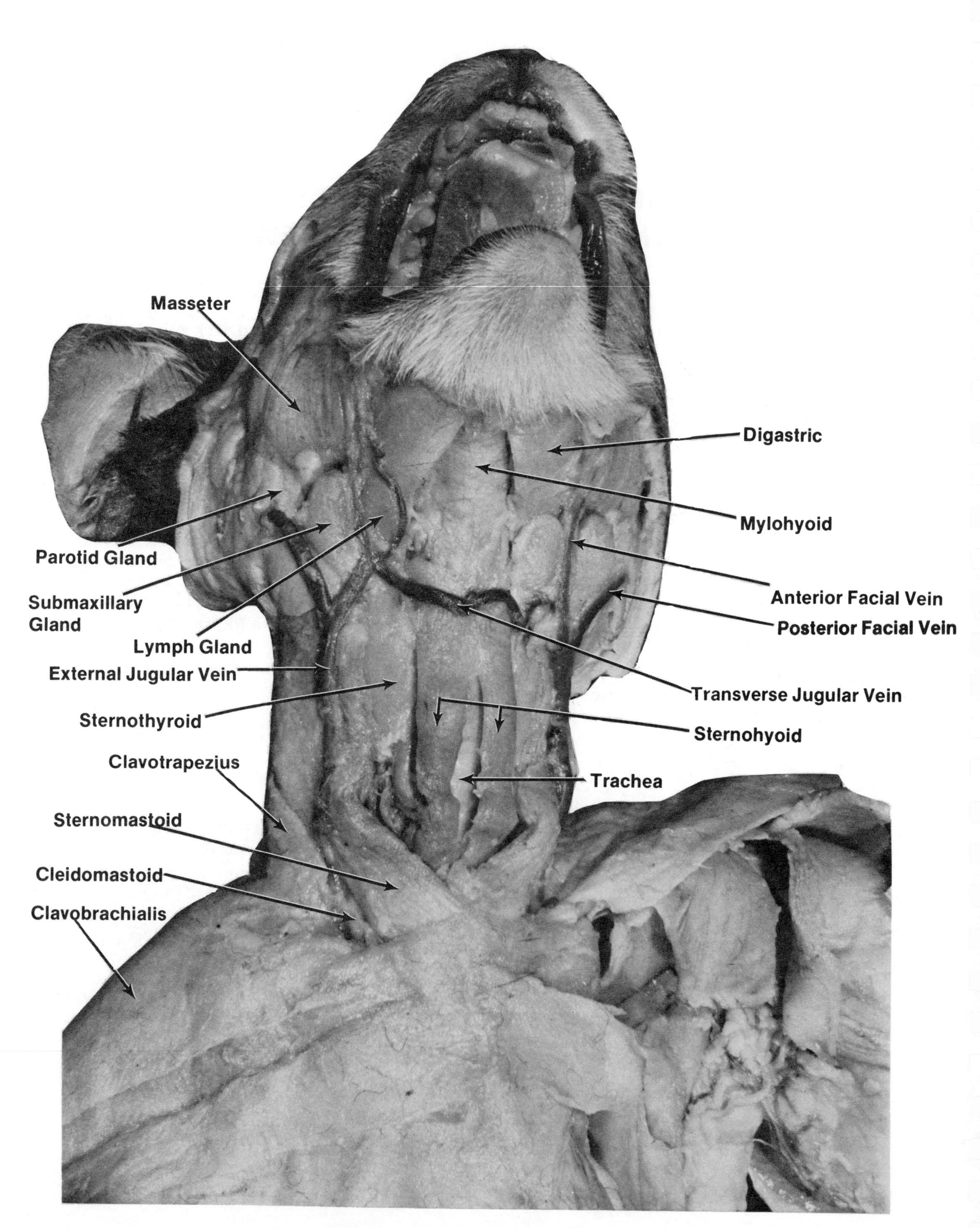


SUPERFICIAL MUSCLES - HEAD, VENTRAL VIEW

# SUPERFICIAL MUSCLES
# HEAD AND NECK, LATERAL VIEW

Turn the cat to expose the lateral surface of the head and neck as in the photo.

Although most of the structures have been seen in the last photo, several are more prominent in the lateral view.

**Masseter** - The most prominent muscle on the lateral surface of the face is the large, rounded masseter. It is the primary muscle used for chewing. It extends from the zygomatic bone to the lateral surface of the mandible. Touch your cheek and gnash your teeth. You will feel this muscle contracting.

**Temporalis** - Another muscle used in chewing is located immediately dorsal to the masseter. It is the temporalis which originates on the sides of the skull, from the parietal and temporal bones, to insert on the mandible. Touch your temple and gnash your teeth. You will feel this muscle contracting.

The *salivary glands* will be discussed in a later section (see page 72).

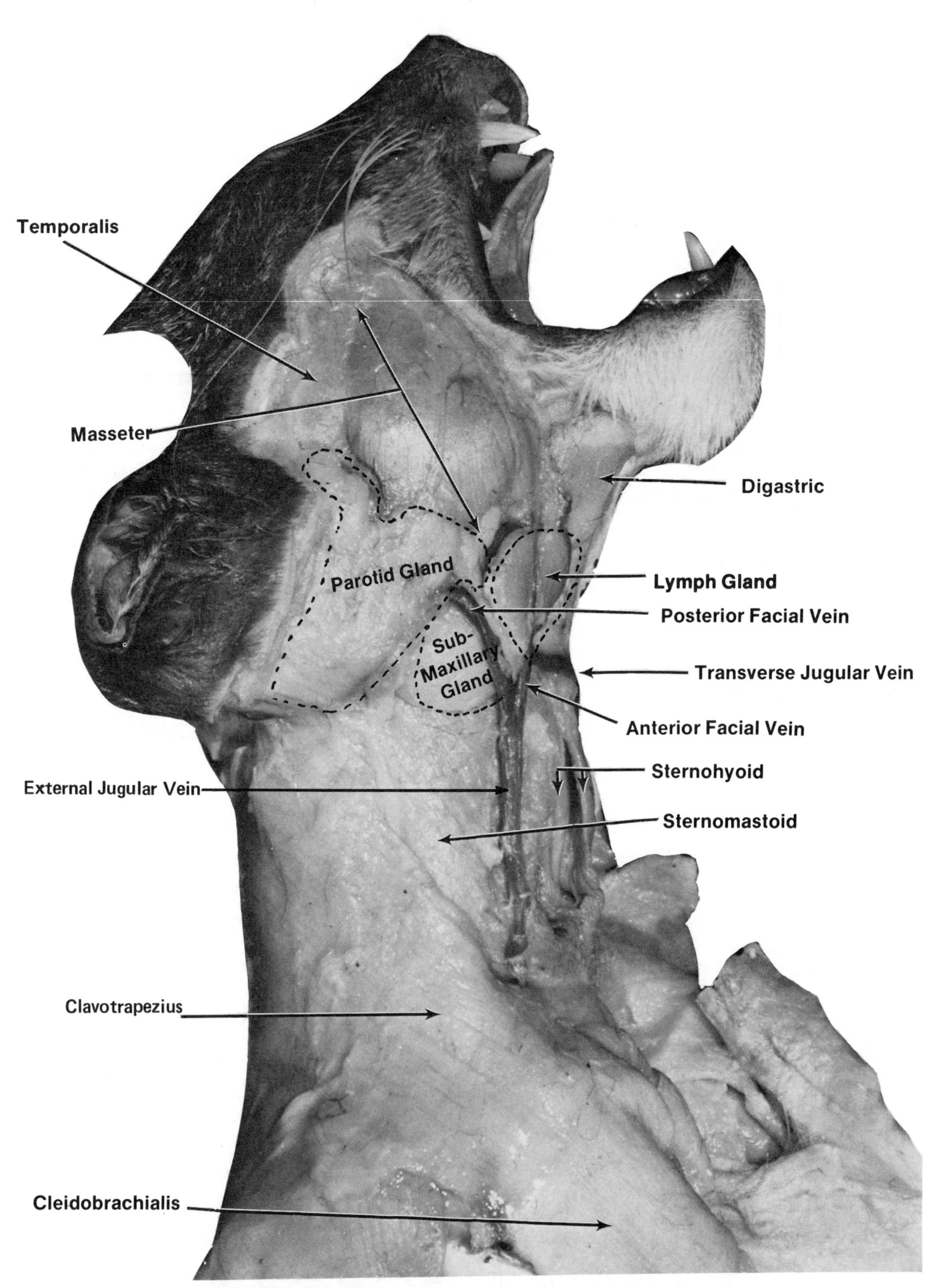


SUPERFICIAL MUSCLES - HEAD, LATERAL VIEW

# SUPERFICIAL MUSCLES - FORELIMB, MEDIAL VIEW

Examine the entire forelimb. Separate the superficial musculature and study the two accompanying photos.

**Clavobrachialis** (Clavodeltoid) - This is the most anterior muscle in the shoulder and arm region. It originates at the clavicle, crosses the shoulder and upper arm, and inserts upon the ulna near the elbow. It is a continuation of the *clavotrapezius* which crosses the neck from the mid-dorsal region and ends at the clavicle. The clavicle is thus the insertion of the first muscle and the insertion of the second. The two muscles work together to extend the humerus, turn the head, and flex the elbow.

Most of the superficial muscles covering the humerus have been described earlier (see pages 21 and 23. They include the: pectoantebrachialis
pectoralis major
pectoralis minor
latissimus dorsi

**Epitrochlearis** - This is a thin supeficial muscle lying upon the medial surface of the humerus. It originates from the ventral border of the latissimus dorsi. It inserts by a thin aponeurosis upon the lower forelimb. It acts together with the triceps brachii an extensor of the elbow. It is not found in humans.

**Triceps Brachii** - This muscle has three heads in both the cat and in man. Only the *long head* of the muscle is partially visible in the accompanying photo. Most of the muscle is hidden by the epitrochlearis. The triceps brachii acts to extend the forearm. It will be more fully described when the cat is examined in dorsal view (see page 39).

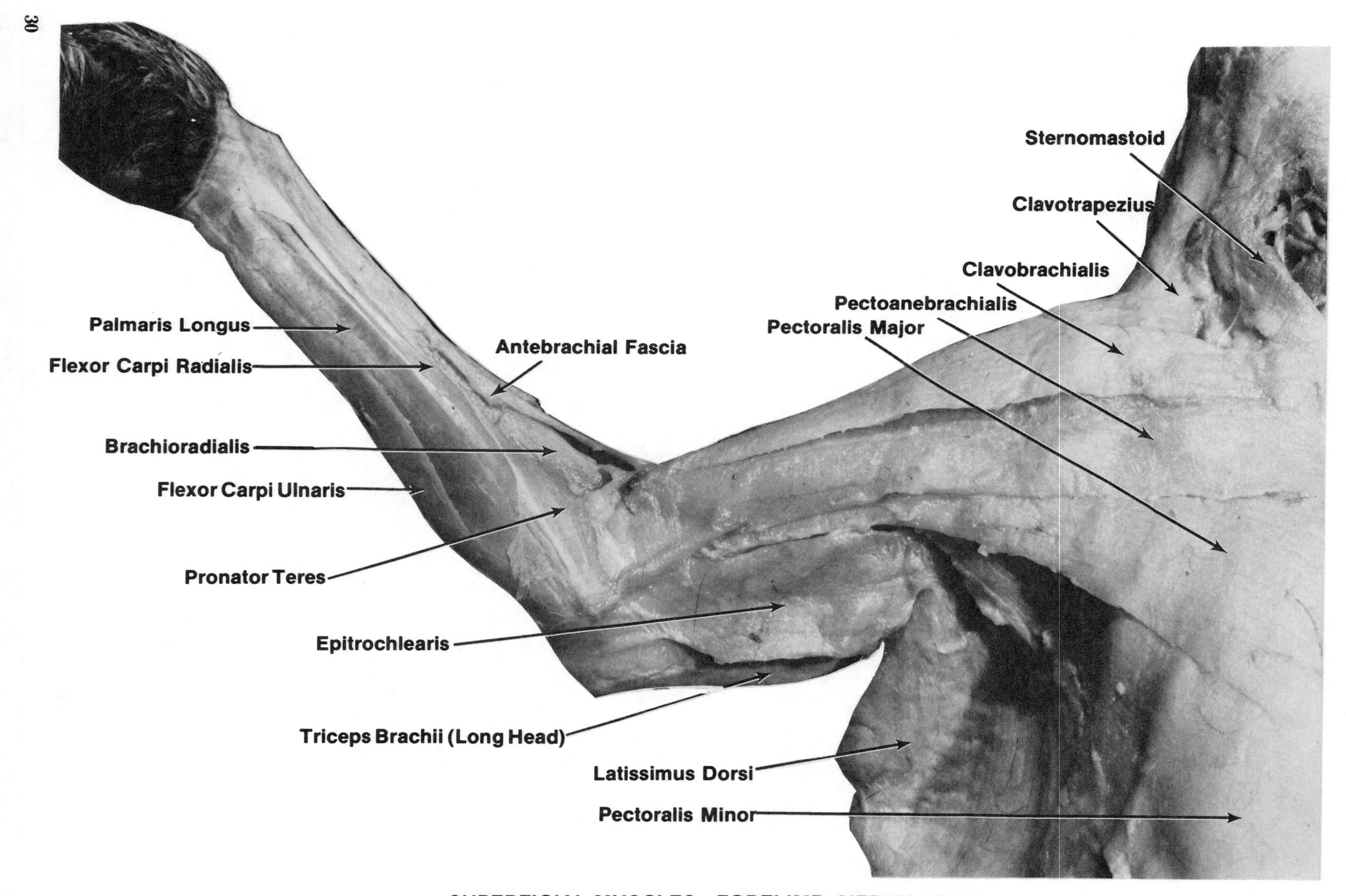


SUPERFICIAL MUSCLES - FORELIMB, MEDIAL VIEW

# SUPERFICIAL MUSCLES
# LOWER FORELIMB, MEDIAL VIEW

The muscles of the lower forelimbs generally originate from the humerus and insert upon the carpals or digits. Those on the medial surface function to flex the wrist and digits.

These muscles are seen in the photo. They include the:

**Flexor Carpi Ulnaris**
**Palmaris Longus**
**Brachioradialis** (See description of this muscle, page 41.)
**Flexor Digitorum Profundus**
**Flexor Carpi Radialis**
**Pronator Teres** - This muscle is also seen in the photo. As its name indicates, it rotates the palm of the hand downward, to the prone position. It originates on the medial epicondyl of the humerus and inserts upon the medial surface of the radius.

A tough layer of fascia, the *antebrachial fascia,* covers these muscles as a sleeve. It is necessary to cut this fascia to expose the muscles. Separate these muscles as in the photo and observe their tendons.

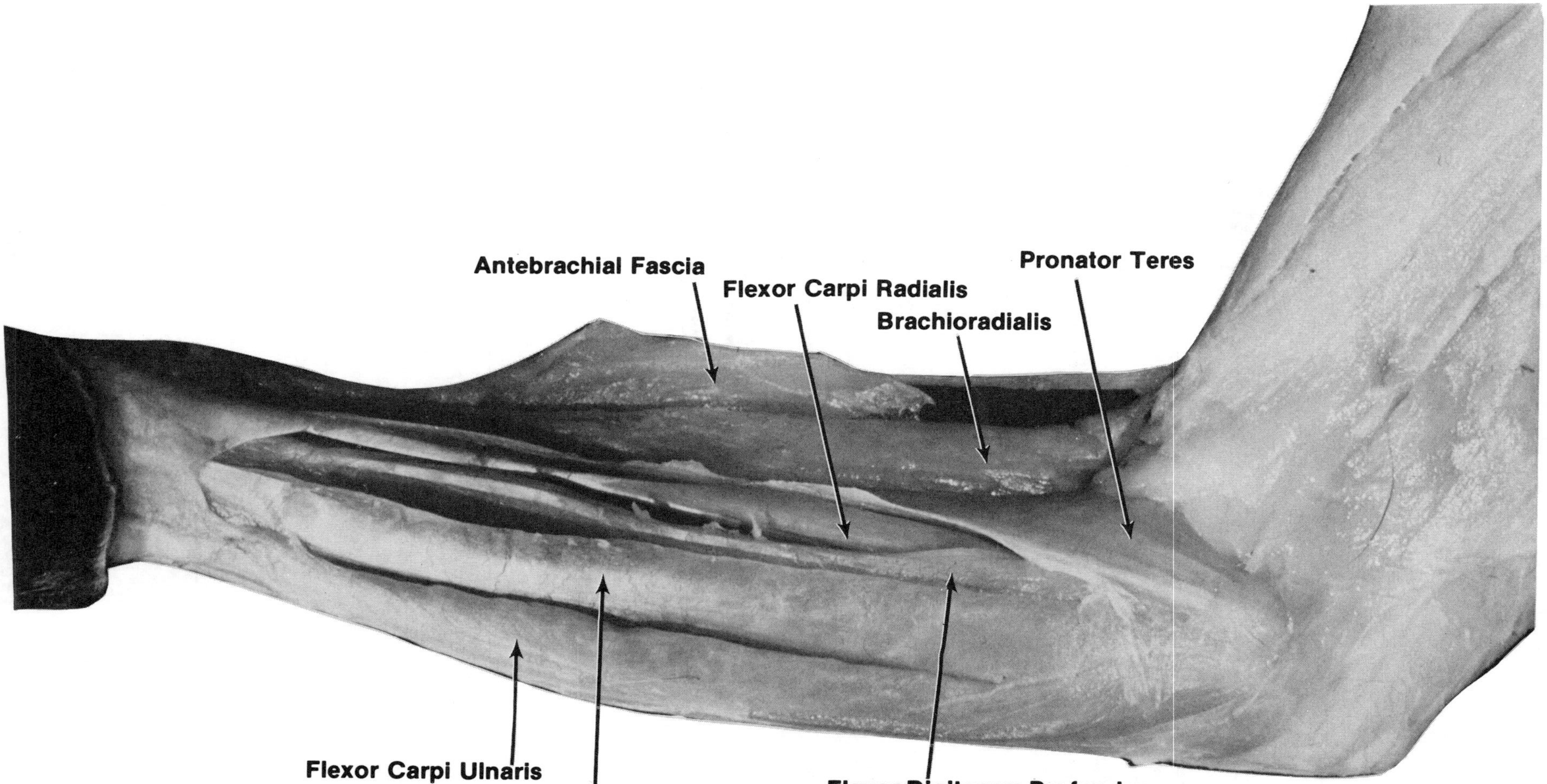

SUPERFICIAL MUSCLES—LOWER FORELIMB, MEDIAL VIEW

# SUPERFICIAL MUSCLES - HIND LIMB, MEDIAL VIEW

Remove the superficial fascia and fat from the medial surface of the thigh. Separate the more obvious muscle masses as in the photo. Do not cut any blood vessels or nerves. The blood vessels seen are the *femoral vein* and *artery* upon the shank. The *femoral nerve* and *saphenous nerve* may also be seen.

Observe the two large superficial muscles on the ventro-medial surface of the thigh.

**Sartorius** - This muscle occupies the anterior half of the thigh. It resembles a flattened band about 1½ inches wide. Separate it from the neighboring muscles. It originates from the crest of the ilium and inserts upon the tibia and patella. It adducts and rotates the thigh and extends the knee. A part of it extends over the thigh to the lateral surface.

**Gracilis** - The second medial superficial thigh muscle is also broad and thin. It covers the posterior portion of the medial (inner) thigh. Its origin is near the pubic symphasis and it inserts upon the medial surface of the tibia. It is an adductor of the thigh.

There are several muscles visible between the sartorius and the gracilis. These include the:

**Adductor Femoris** - A portion of this muscle is seen immediately anterior to the gracilis, while the major part lies deep to the gracilis. Its name clearly indicates its action. It originates from the pubis and inserts on the femur.

**Adductor Longus** - This narrow muscle also belongs to the adductors of the thigh. Its origin and insertion are similar to those of the adductor femoris.

**Pectineus** - This is another adductor of the thigh. It is a narrow muscle, smaller than the adductor longus. It lies just posterior to the femoral artery and vein. Its origin is on the anterior pubis and its insertion is on the shaft of the femur near the lesser trochanter.

**Iliopsoas** - Only a small portion of the muscle is seen near its point of insertion on the femur. It is an extensive muscle arising from the lumbar vertebrae and from the ilium. It forms a major portion of the dorsal abdominal wall, and is the chief flexor of the trunk. It is homologous to two muscles in man, the *iliacus* and the *psoas major.*

Beneath the sartorius lie four muscles of the anterior thigh collectively known as the *quadriceps femoris.* They will be described when the deeper muscles of the hind leg are considered (see pages 56, 57).

We are now ready to consider the superficial musculature of the lower hind leg.

**Gastrocnemius** - This is the major muscle of the calf. It is visible on both the medial surface as in the photo and the lateral surface. It originates as two separate heads, the *lateral* and the *medial head,* on the femur and inserts upon the heel bone, the *calcaneous,* by way of the long, tough *Achilles tendon.* It acts as an extensor of the foot.

**Soleus** - It lies deep to the gastrocnemius but is also visible superficially. It originates on the fibula and together with the gastrocnemius inserts on the calcaneus by way of the Achilles tendon. It also acts to extend the foot.

**Tibialis Anterior** - This is the superficial muscle covering the anterior portion of the lower hind leg. As its name indicates, it lies upon the tibia. It originates from the tibia and fibula and inserts by means of a long tendon, which crosses the ankle obliquely, upon the first metatarsal. It extends the foot and also inverts it.

**Flexor Digitorum Longus** - This muscle lies on the medial aspect of the lower hindleg, next to the tibia. It arises from the tibia and fibula and inserts upon the terminal phalanges by means of four tendons,one for each digit. It acts as a flexor of the digits.

---

**In the space below make a list of all the superficial muscles you have found on the ventral side of your cat.**

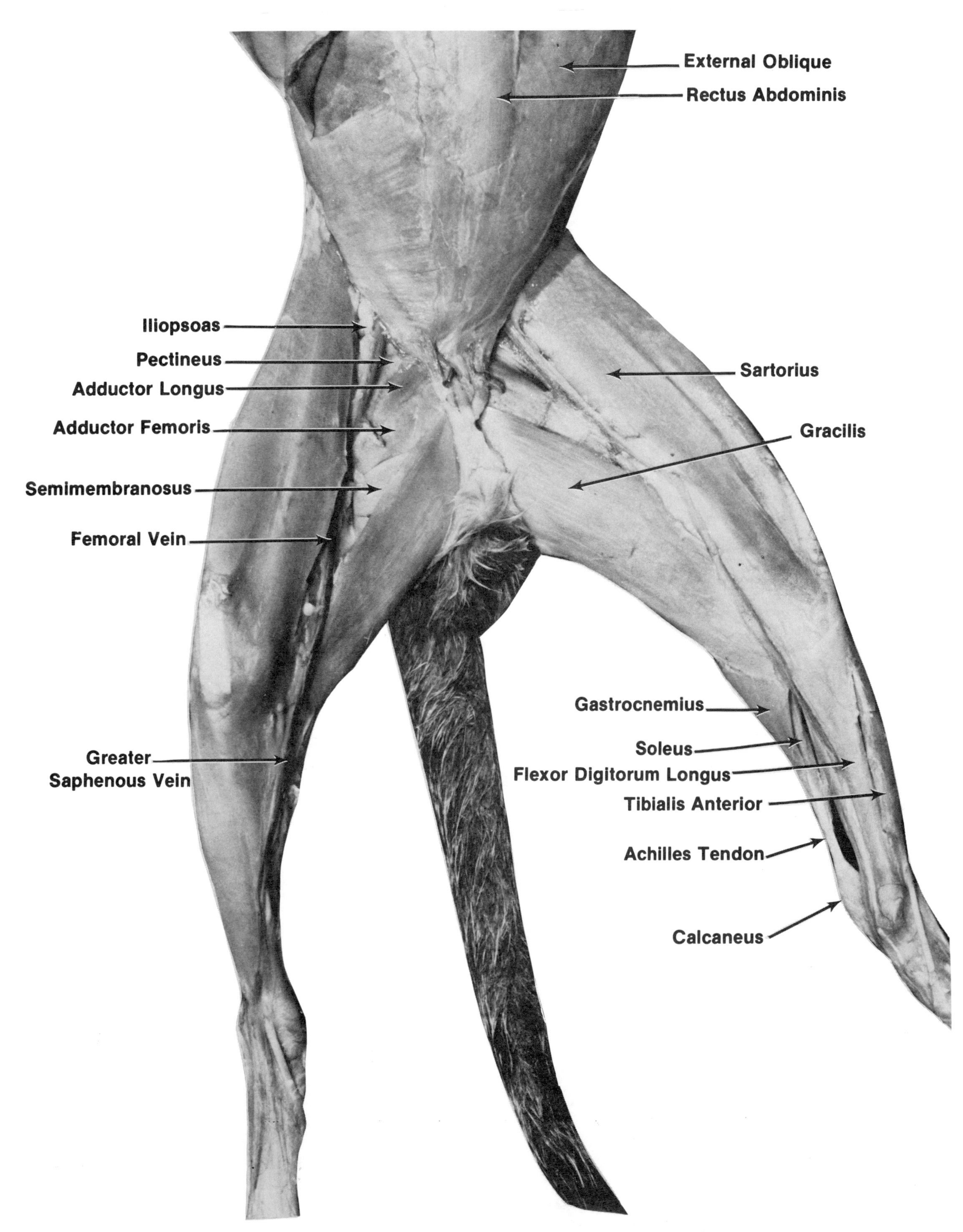


SUPERFICIAL MUSCLES - HIND LIMB, MEDIAL VIEW

# REVIEW
# SUPERFICIAL MUSCLES - ENTIRE BODY, VENTRAL VIEW

You are now ready to *review* all of the superficial muscles on the ventral surface of the body.

Unlike the photos of the previous pages which featured views of different sections of the body, this photo reveals the entire ventral surface.

By this time all of the muscles in the photo should be located, separated, and identified; their *origins*, *insertions*, and *actions*, noted.

Each of the muscles labeled on the photo is numbered. Next to the numbers in the list below enter the name of the muscle that corresponds to that number.

1. ____________________
2. ____________________
3. ____________________
4. ____________________
5. ____________________
6. ____________________
7. ____________________
8. ____________________
9. ____________________
10. ____________________
11. ____________________
12. ____________________
13. ____________________
14. ____________________
15. ____________________
16. ____________________
17. ____________________
18. ____________________
19. ____________________
20. ____________________

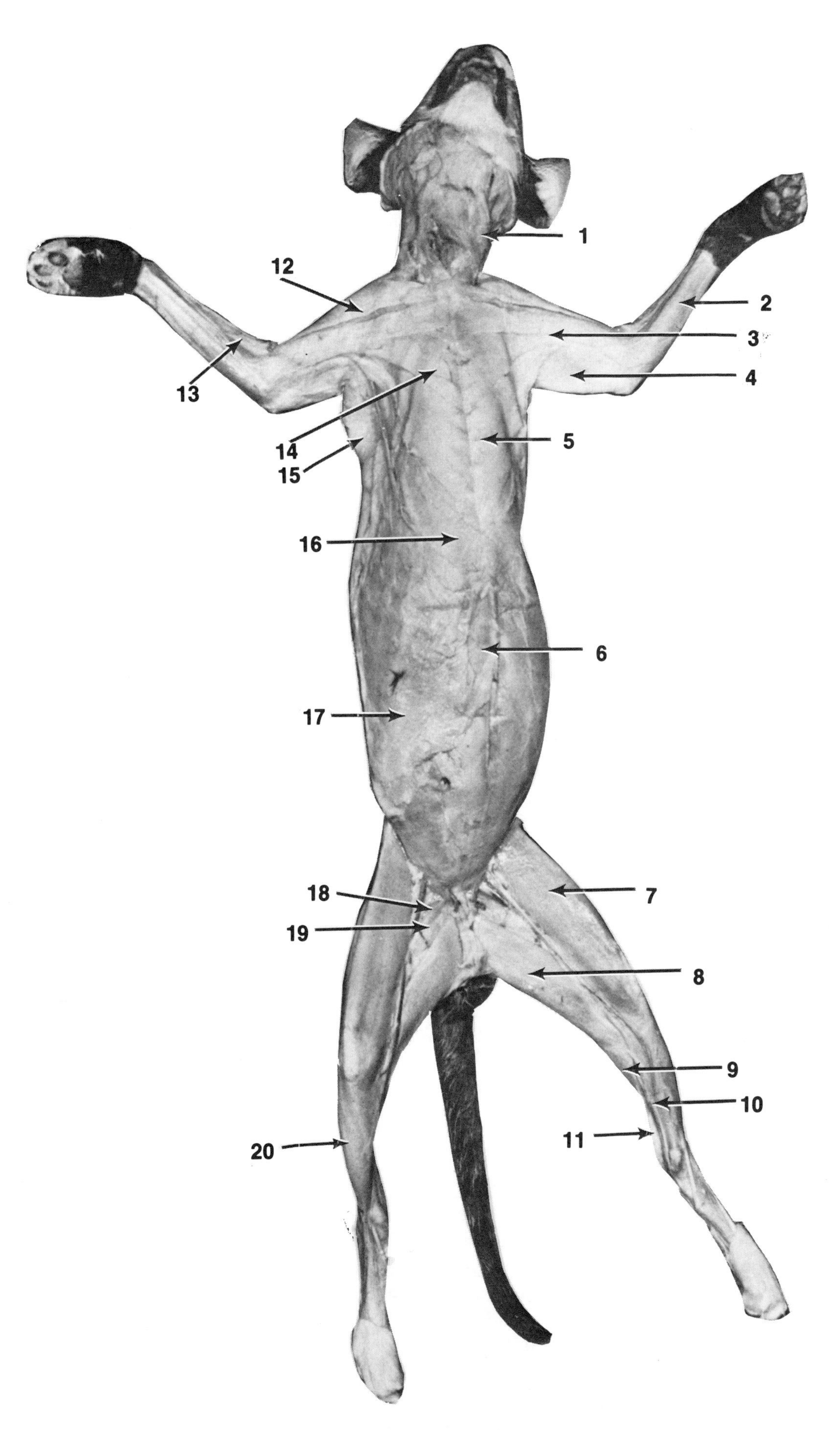


SUPERFICIAL MUSCLES - ENTIRE BODY, VENTRAL VIEW

# SUPERFICIAL MUSCLES - NECK, SHOULDER, AND THORAX, DORSAL VIEW

Turn the cat to expose the dorsal surface of the neck, shoulder, and thorax.
Note: A view of the entire superficial dorsal musculature is seen in the photo on page 46.

In this region the predominant superficial dorsal muscles are those of the *trapezius* group. In man they are combined into a single muscle, while in the cat three different muscles are found. They move the pectoral girdle and the head.

**Clavotrapezius** - This is the most anterior of the three. It covers the dorsal and lateral surfaces of the neck and part of the shoulder. It arised from the back of the head and mid-dorsal line of the neck. It inserts on the clavicle. Here, its fibers are continuous with those of the *clavobrachialis* which originates from the clavicle and continues to the arm. This is clearly seen in the photo.

**Acromiotrapezius** - This is the second muscle of the trapezius group. It is thin, flat, and triangular in shape. It lies posterior to the clavotrapezius and covers most of the scapula and upper thorax. Its origins are the posterior cervical and anterior thoracic vertebrae. It inserts on the spine of the scapula.

**Spinotrapezius** - The third muscle of this group lies posterior to the acromiotrapezius. It originates from the spinous processes of the posterior thoracic vertebrae. It is also broad, thin, and triangular. It too inserts upon the spine of the scapula.

**Latissimus Dorsi** - The remainder of the back of the cat, the posterior portion, is covered by this large muscle. Although, as its name indicates, it is primarily a muscle of the dorsal surface, it was described earlier with the ventral musculature (see page 23). This is due to the fact that its fibers extend laterally to insert upon the humerus, on the ventral surface.

Another group of dorsal muscles is the *deltoid.* As was true for the trapezius, it is a single muscle in man but three different muscles in the cat. They act together to raise (adduct) the humerus and to rotate it. They include the following:

**Clavobrachialis** (Clavodeltoid) - This flat, thin muscle which covers the anterior portion of the shoulder and arm is a continuation of the clavotrapezius. It originates in the clavicle and inserts into the ulna near the elbow. Although it has been described earlier with the muscles of the upper limb (see page 29), it is properly considered one of the deltoid group, as its secondary name clearly indicates.

**Acromiodeltoid** - This short, thick muscle originates at the acromion process of the scapula. It inserts on the proximal end of the humerus.

**Spinodeltoid** - This is the most posterior muscle of the deltoid group. It originates on the spine of the scapula, to which it runs parallel, and inserts on the humerus. It is readily identified by its rectangular shape and its relationship to the surrounding muscles.

**Levator Scapulae Ventralis** - Here we have another muscle named for its action. It draws the scapulae anteriorly. It is a this band of muscle whilch extends caudally from the first cervical vertebra, the atlas, and from the occipital bone to insert on the scapula, passing beneath the clavotrapezius. It is visible superficially between the clavotrapezius and the acromiotrapezius. This muscle is not present in man.

**Triceps Brachii** - Clearly visible on the lateral surface of the forelimb are two of the heads of this large muscle, the *long head* and the *lateral head*. In the medial view which was described earlier, almost the entire muscle was hidden by another muscle, the epitrochlearis. The third head, the *medial head,* lies deep to the other two heads.

**Long Head** - This segment is the largest of the three heads. It is a spindle-shaped mass on the dorso-lateral aspect of the forelimb. It arises from the scapula.

**Lateral Head** - This flattened segment lies upon the lateral surface of the upper forelimb, as its name indicates. It originates from the dorsal surface of the humerus.

**Medial Head** - This deeper layer also arises from the dorsal surface of the humerus. It is not seen in the photo. It can be seen in later photos (pages 53 and 62) when the deeper muscles of the forelimb are exposed.

All three heads of the triceps brachii insert in common upon the *olecranon process* of the ulna, and act together to extend the forelimb.

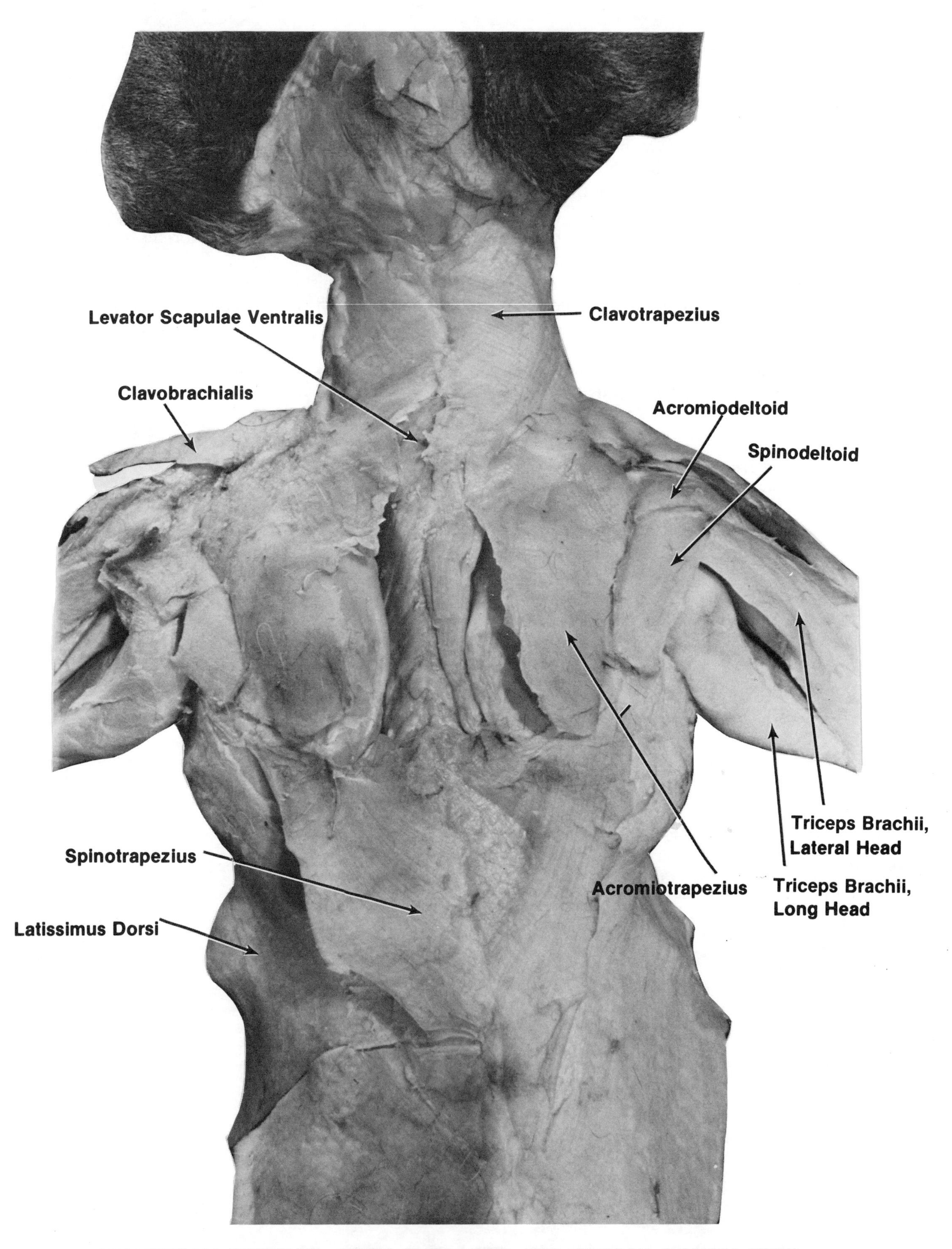


SUPERFICIAL MUSCLES - NECK, SHOULDER, AND THORAX, DORSAL VIEW

# SUPERFICIAL MUSCLES - FORELIMB, LATERAL VIEW

In the shoulder area adjacent to the forelimb several muscles are seen. These have been described in the last section. They include the:

clavobrachialis
acromiodeltoid
spinodeltoid

In the upper forearm we recognize the long head and the lateral head of the triceps brachii, also described in the last section.

**Brachialis** - This muscle, not seen previously, lies anterior to the lateral head of the triceps brachii. Only a very small portion is visible here. It is seen in its entirety only when the lateral head of the triceps brachii is transected and reflected (see the photo on page 62). It arises from the lateral surface of the humerus and inserts upon the proximal end of the ulna. It is a flexor of the elbow.

We are now ready to examine the muscles of the lower forelimb.

A tough layer of fascia, the *antebrachial fascia,* covers the muscles of the lower forelimb. This must first be cut. Observe the tendons of these muscles at the wrist. Separate the superficial muscles. These long, thin muscles cover the lower forelimb and generally originate from the humerus. Their action is to extend the wrist and digits and to supinate the foot.

**Brachioradialis** - This is the most anterior of these muscles. Its origin is the middle of the humerus, its insertion at the distal end of the radius. It rotates the radius and supinates the paw.

**Extensor Carpi Radialis Longus** - This muscle inserts upon the second metacarpal. Its action is to extend the forefoot. Another muscle, the *Extensor Carpi Radialis Brevis* is usually partly covered by the first muscle (Longus), and is often fused with it. It inserts on the third metacarpal.

**Extensor Digitorum Communis** - This is a long muscle whose tendon divides at the wrist and attaches to the second, third, fourth, and fifth digits, all of which it extends.

**Extensor Digitorum Lateralis** - It is also a long muscle of the lower forelimb. It inserts upon the third, fourth, and fifth digits, and extends them.

**Extensor Carpi Ulnaris** - It inserts on the fifth metacarpal.

All of these muscles act to extend the lower forelimb, the wrist, or the paw.

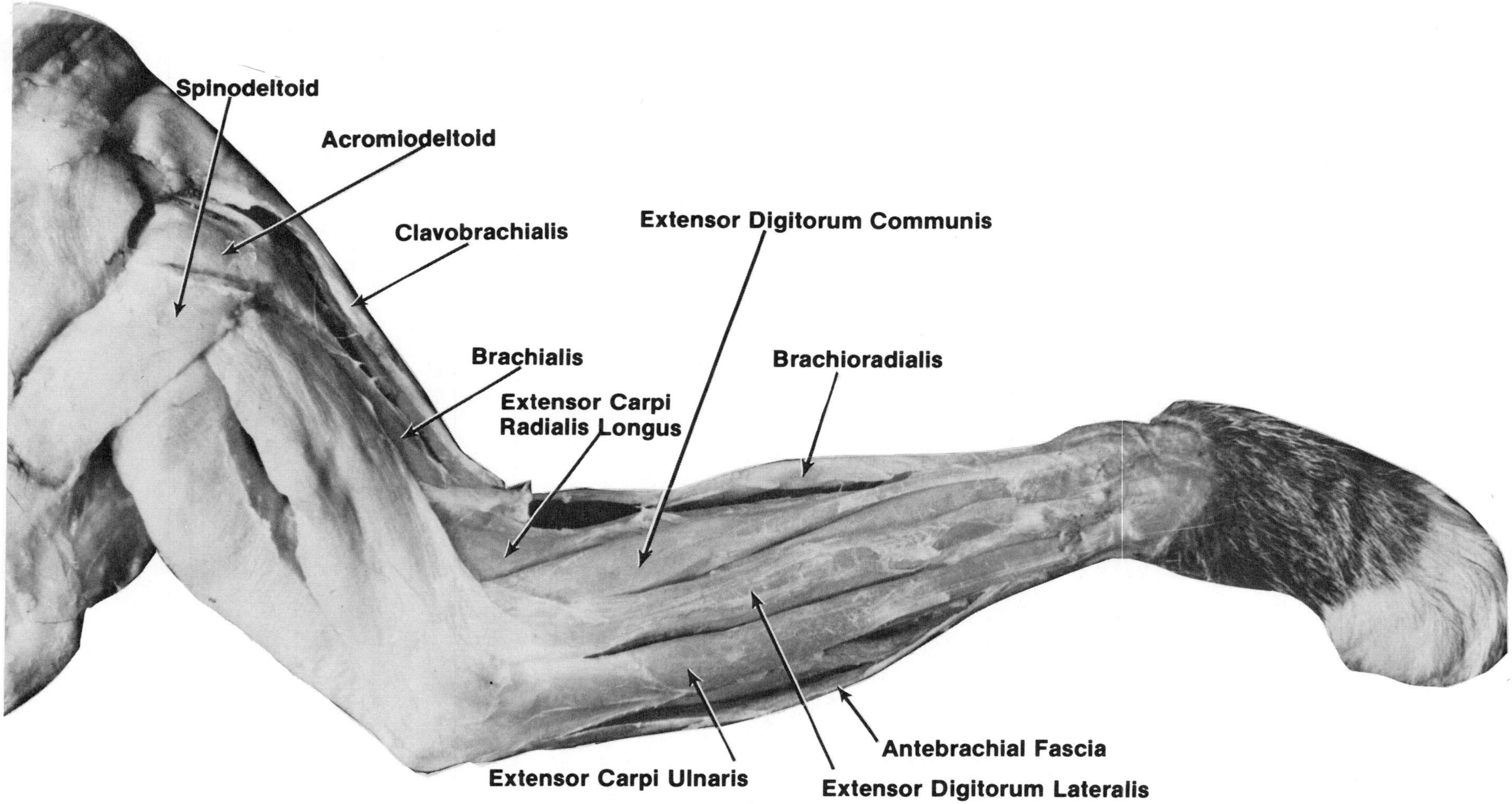

Spinodeltoid

Acromiodeltoid

Clavobrachialis

Extensor Digitorum Communis

Brachialis

Brachioradialis

Extensor Carpi Radialis Longus

Antebrachial Fascia

Extensor Carpi Ulnaris

Extensor Digitorum Lateralis

SUPERFICIAL MUSCLES - FORELIMB, LATERAL VIEW

# SUPERFICIAL MUSCLES
# HIPS AND THIGH, DORSAL VIEW

Examine the area of the hip and thigh. Separate the muscles as seen in the photo.

**Tensor Fascia Lata** - This large, fan-shaped muscle lies on the proximal portion of the lateral surface of the thigh. It inserts upon the *fascia lata,* a tough white band of connective tissue. Its origin is on the ilium. It acts to tense the fascia lata and to extend the thigh.

Three *gluteal muscles* are visible on the surface.

**Gluteus Medius** - This is a thick, short muscle almost entirely covered by the tensor fascia lata. It originates from the ilium and the transverse processes of sacral and caudal vertebrae. Its insertion is upon the greater trochanter of the femur. It acts as an abductor of the thigh.

**Gluteus Maximus** - In the cat this is a relatively small muscle, while in humans it forms the primary tissue of the buttocks. Its fibers run laterally from sacral and caudal vertebrae and insert upon the proximal femur. It abducts the thigh.

**Caudofemoralis** - This muscle lies posterior to the gluteus maximus. In humans it is united with the gluteus maximus. It originates on the caudal vertebrae and inserts on the patella by way of a long narrow tendon. It abducts the thigh and extends the leg at the knee.

Another group of muscles upon the posterior thigh is the *hamstrings.* As in humans, this group includes three muscles:

**Biceps Femoris** - This is a large and powerful muscle covering much of the lateral surface of the thigh. It is 2-2½ inches wide throughout most of its length. It extends from the ischium to the tibia. Much fat is usually deposited under its lower end. When this fat is removed, the large *sciatic nerve* is exposed. The area occupied by the fat is the *popliteal space.* The *popliteal artery* and *vein* as well as a lymph gland are embedded within the fat. Its action is to flex the shank and to abduct the thigh.

**Semiteninosus** - This muscle is located medial to the biceps femoris, lying on the posterior aspect of the thigh. It arises from the ischium and inserts by a tendon on the tibia. It acts to flex the knee. This muscle is also described when the deeper muscles of the thigh are examined (see page 56 and the photos on pages 58 and 67).

**Semimembranosus** - This large muscle is deep to the gracilis and medial to the semitendinosus. It too originates from the ischium while its insertion is on the distal end of the femur. It acts to extend the thigh.

The relationship between the semimembranosus and the semitendinosus is best seen when the biceps femoris has been transected (see the photo on page 67).

An extensive aponeurosis, the *lumbodorsal fascia,* covers the lower back to join the superficial muscles in that area.

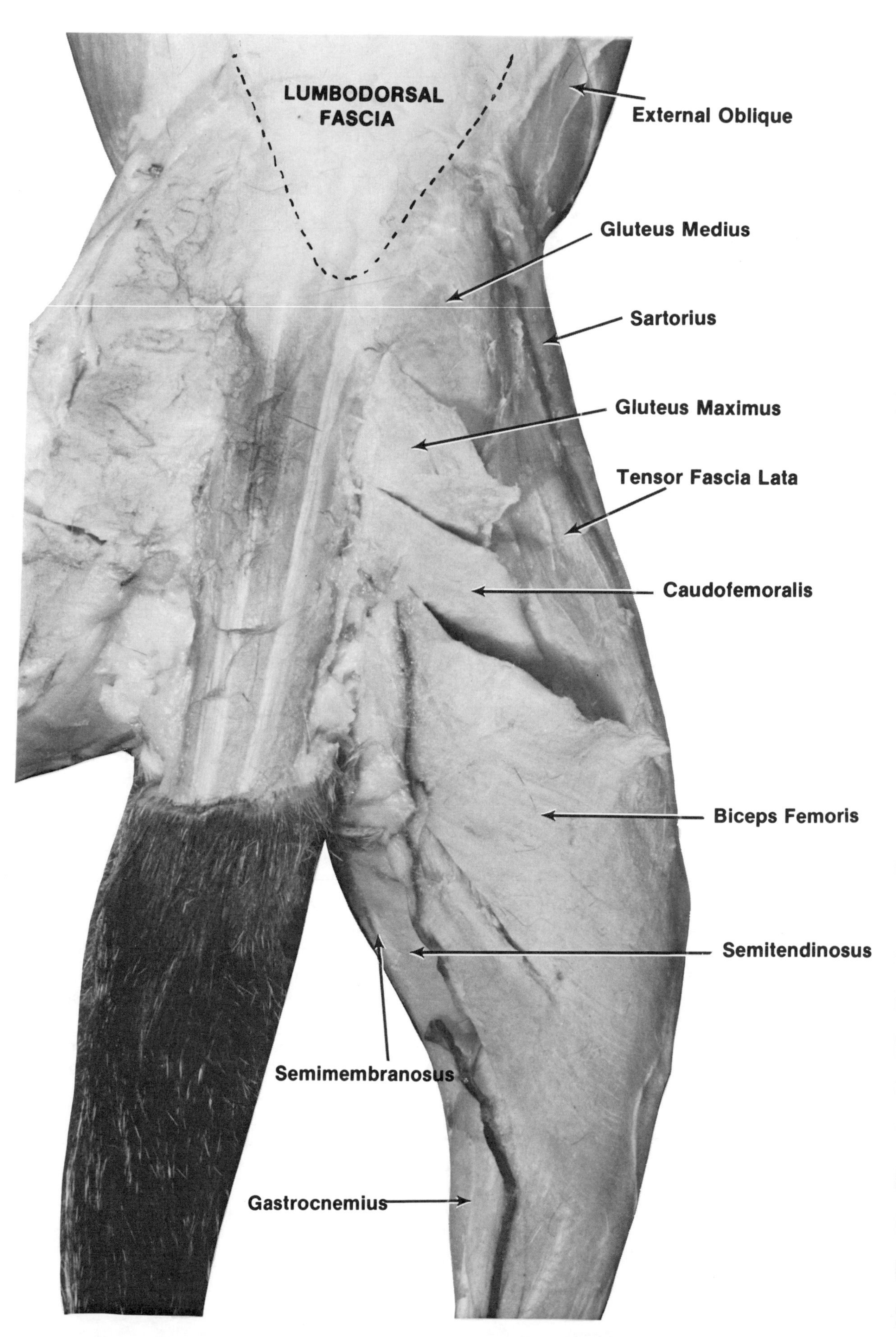


SUPERFICIAL MUSCLES - HIPS AND THIGH, DORSAL VIEW

# REVIEW
# SUPERFICIAL MUSCLES - ENTIRE BODY, DORSAL VIEW

You are now ready to *review* all of the superficial muscles on the dorsal surface of the body.

The photo reveals the entire dorsal surface. However, none of the muscles has been separated. By this time you should be able to name all of the muscles indicated. Then, draw directly upon the photograph the limits and extent of each muscle. This will indicate how well you have done the earlier parts of the dissection.

Each of the muscles labeled on the photo is numbered. Next to the numbers in the list below enter the name of the muscle that corresponds to that number.

1. ______________________
2. ______________________
3. ______________________
4. ______________________
5. ______________________
6. ______________________
7. ______________________
8. ______________________
9. ______________________
10. ______________________
11. ______________________
12. ______________________
13. ______________________
14. ______________________
15. ______________________
16. ______________________
17. ______________________
18. ______________________
19. ______________________
20. ______________________

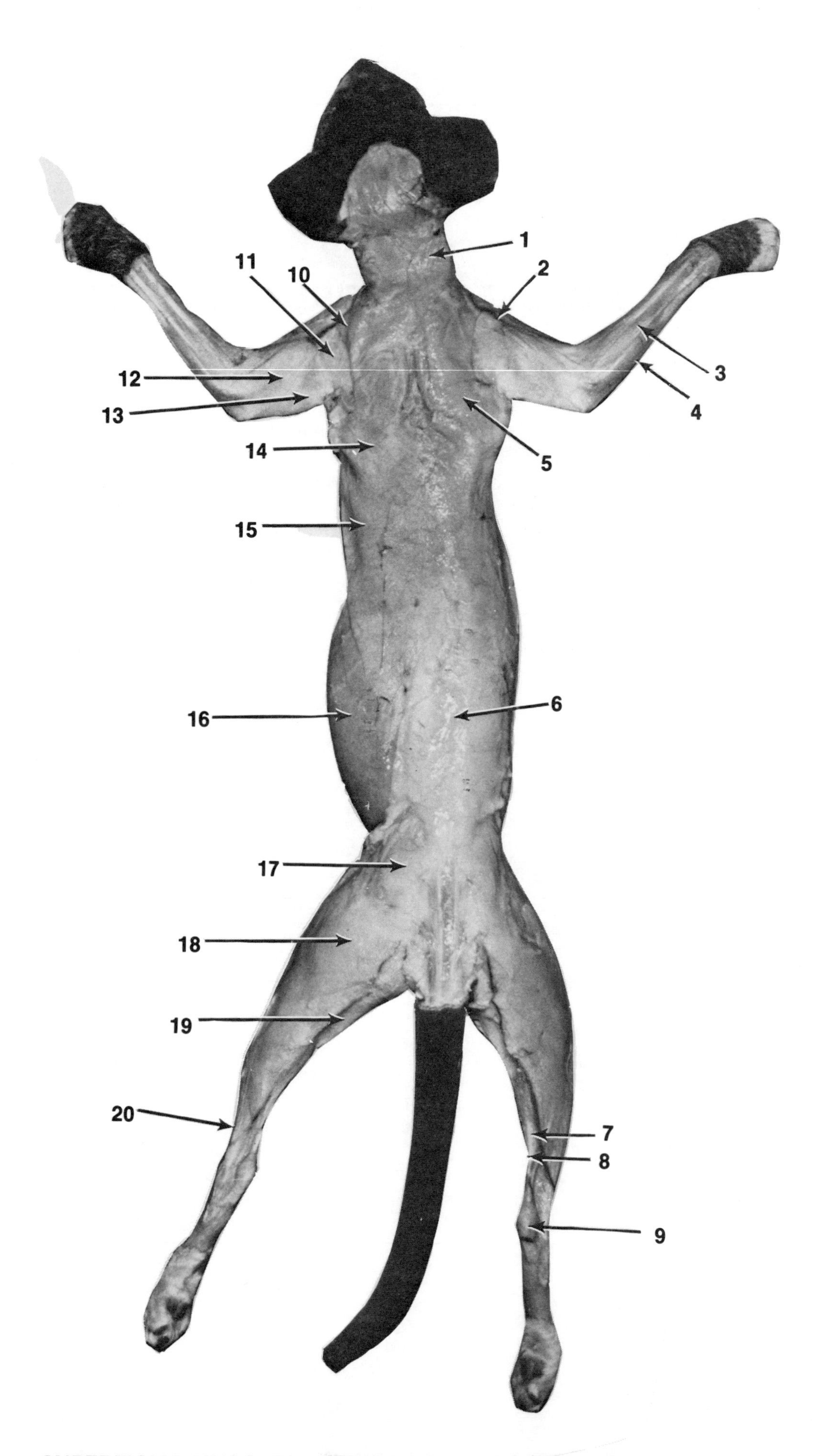


SUPERFICIAL MUSCLES – ENTIRE BODY – DORSAL VIEW

# DEEP MUSCLES - INTRODUCTION

The dissection of the deeper muscles should be done on one side only, leaving the superficial muscles on the second side intact. Compare the superficial muscles on one side with the deeper muscles on the other side. This procedure permits later utilization of superficial muscles for review and study. Also, if for some reason, the deeper muscles on one side are improperly dissected or destroyed, those of the second side are then still available. Similarly, dissect the deeper muscles of only one forelimb and one hind limb, leaving the superficial muscles intact upon the second limb.

In order to expose the deeper muscles it is best to *transect* each of the superficial muscles. This is done by cutting them at right angles to the direction of the fibers at the central belly area. They are then folded back, or *reflected,* to their origins and insertions. They can thus easily be "reconstructed"at any time in order to examine the relationship of the deeper muscles to those of the superficial layers.

You will find that the deeper muscles are quite different from those above them. The variations in shapes, lengths, directions of the fibers, as well as their actions are revealed only after they have been exposed. It is impossible to guess about them by viewing the superficial muscles alone. For example, could we predict the unusual shape of the serratus ventralis muscle, seen in the next photo by merely viewing the superficial thoracic musculature? Such examples abound when considering the entire cat muscalature.

The description of the deeper muscles will repeat the pattern followed for the superficial muscles; first the anterior ventral area, then the posterior, finally the deeper muscles of the dorsal region. A Self-Quiz for students is found at the end of the entire unit on muscles.

# DEEP MUSCLES
# THORAX AND SHOULDER, VENTRAL VIEW

Transect the following large muscles on the left side:

pectoantebrachialis
pectralis major
pectoralis minor
xiphihumeralis

Your cat should now appear as in the photo.

Identify the following muscles:

**Serratus Ventralis** - This is a large fan-shaped muscle. Its name is derived from the saw tooth-like edges of the muscle strips. It arises by a number of slips from the lateral surface of the upper eight or nine ribs and from the transverse processes of the last five cervical vertebrae to insert upon the dorsal margin of the scapula ventral to the rhomboideus. In quadrupeds it forms, together with the pectoralis, a muscular sling that transfers much of the weight of the body to the pectoral girdle and appendages.

That portion of the serratus ventralis which arises from the cervical vertebrae is at times termed *levator scapulae* although it is not a separate muscle as in man. It draws the scapula ventrally and anteriorly. In humans the *serratus anterior* is homologous to the serratus ventralis.

**Scalenus** - This is a long muscle that extends longitudinally along the ventro-lateral surface of the neck and thorax. It is divisible into three separate muscles; the *scalenus anterior, medius,* and *posterior.*

The largest and most readily identifiable of the three is the scalenus medius. You can readily identify the scalenus medius and posterior in the photo. The anterior segment is continuous with the *transversus costarum* muscle. The entire scalenus group originates on the ribs and inserts upon the transverse processes of the cervical vertebrae. They bend the neck and draw the ribs anteriorly.

**Transversus Costarum** - This muscle is not found in humans. It is located near the mid-ventral line where it crosses diagonally from its origin on the sternum to its insertion on the first rib. It acts together with the scalenus. As seen in the photo, it covers the anterior end of the rectus abdominis muscle.

**Rectus Abdominis** - This muscle was described earlier when the superficial abdominal muscles were discussed. It extends, however, anteriorly to the thorax as well. It originates at the upper ribs and sternum along the mid-ventral line, as seen in the photo. It extends posteriorly, parallel to its partner on the right and left mid-ventral line, to the pubis.

On the ventral surface of the scapula two muscles are clearly seen.

**Subscapularis** - This large, flat muscle occupies the *subscapular fossa* of the scapula. It inserts on the humerus to adduct it and to rotate it medially.

**Teres Major** - This muscle originates upon and covers the axillary and posterior borders of the scapula. Its insertion is upon the humerus by means of a tendon in common with the latissimus dorsi. Its action is to rotate and flex the humerus.

**Teres Minor** - This muscle is not seen in ventral view. It will be described later when discussing the deep muscles of the dorsal surface (see page 59).

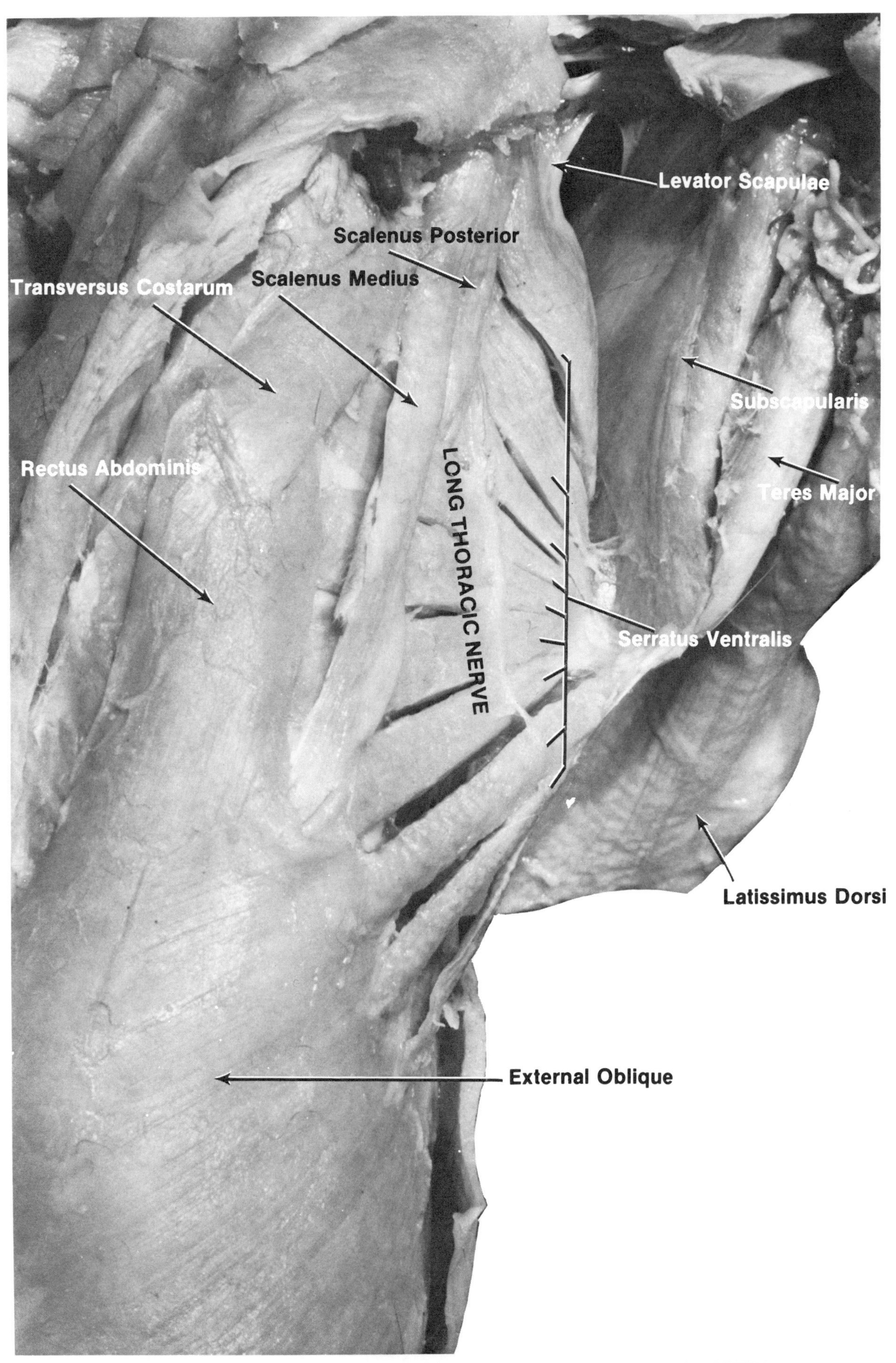


DEEP MUSCLES - THORAX AND SHOULDER, VENTRAL VIEW

# DEEP MUSCLES - CLOSE UP - VENTRAL THORAX

In order to expose the *intercostal muscles,* located between one rib and the next, it is necessary to transect and reflect the rectus abdominis and the external oblique muscles. This has been done in the accompanying photo.

Some of the muscles seen in the previous photo are readily recognized in this close-up view of the thorax and shoulder. These include the:

scalenus medius
transversus costarum
serratus ventralis

The rectus abdominis has been transected and the external oblique reflected. The origin of the transversus costarum at the sternum has been cut and the muscle has been displaced laterally.

The *intercostal muscles* are composed of two sets:

**External Intercostals** - Their fibers are directed in a caudo-ventral direction. As their name indicates, they are the more superficial of the two.

**Internal Intercostals** - Their fibers are directed in a caudo-dorsal direction, at right angles to those of the external intercostals. They are the deeper rib muscles.

The intercostals draw the ribs together during respiration. The rib cage is raised by the external intercostals during inspiration and lowered by the internal intercostals during expiration. The internal intercostals are located along the entire inter-space between the ribs, from dorsal side to sternum, while the external intercostals cover the dorsal and lateral, but not the ventral part of the inter-space.

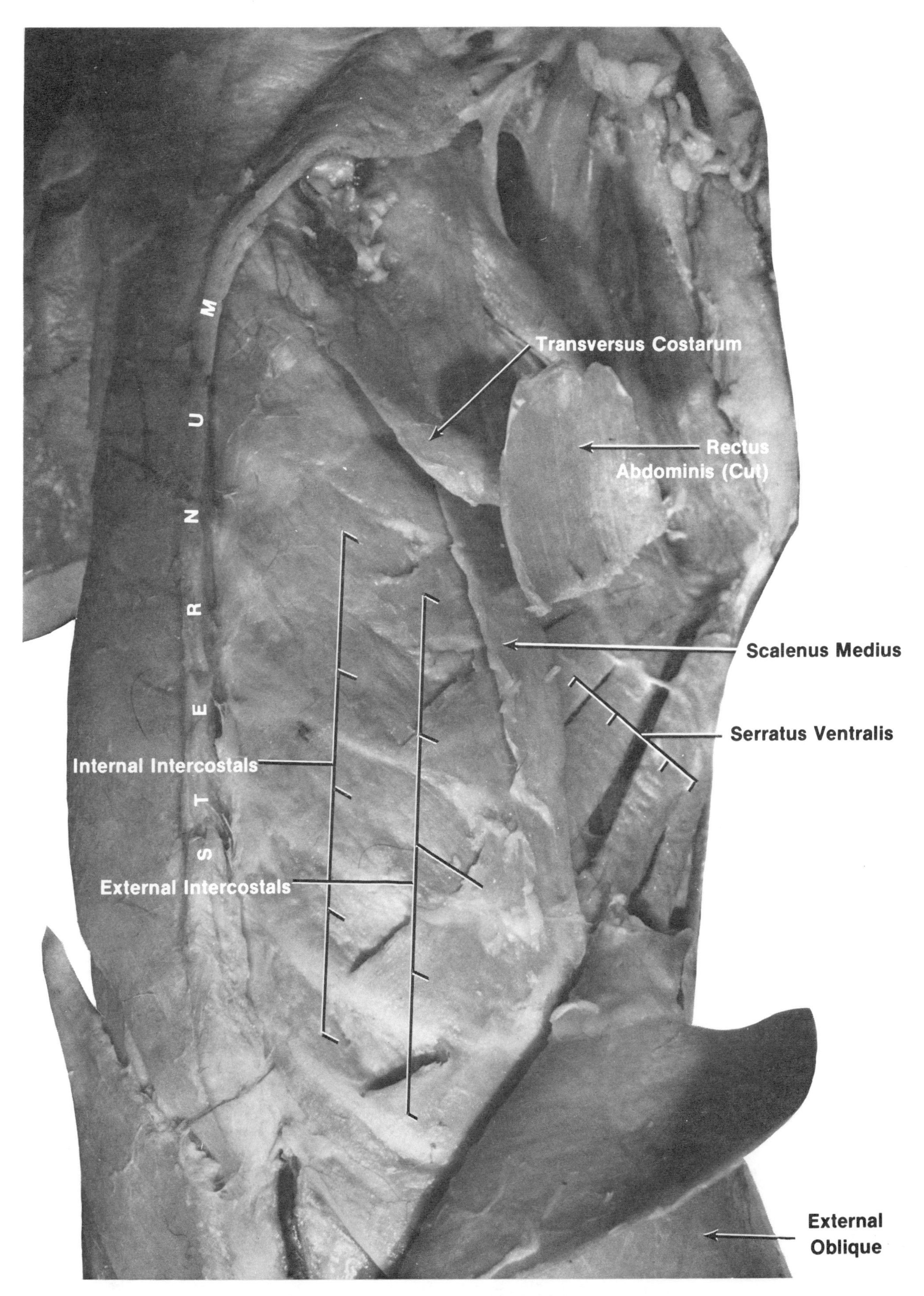


DEEP MUSCLES – CLOSE-UP, VENTRAL THORAX

# DEEP MUSCLES - FORELIMB, MEDIAL VIEW

In order to observe the deeper muscles of the forelimb it is necessary to transect the following superficial muscles:

clavobrachialis
pectoantebrachialis
pectoralis major
pectoralis minor
latissimus dorsi

Several muscles which could not be seen previously now come into view. These include the:

**Biceps Brachii** - This muscle, although superficial and prominent in man, lies deep to the pectoral and brachial muscles of the cat. It is the primary flexor of the antebrachium (lower foreleg). It lies upon the anterio-medial surface of the humerus. It arises from the dorsal edge of the glenoid cavity and inserts upon the radial tuberosity. In humans it arises by two separate tendons, thus its name.

**Triceps Brachii, Medial Head** - This division of the triceps brachii cannot be seen superficially and was therefore not described earlier. As seen in the photo, it lies parallel and medial to the biceps brachii. Its origin is on the dorsal surface of the humerus, its insertion on the olecranon process of the ulna. It acts to extend the forelimb.

**Triceps Brachii, Long Head** - This muscle was only partly visible in superficial view since most of it was covered by the epitrochlearis muscle. In the photo the epitrochlearis has been transected and reflected permitting a full medial view of this large and powerful muscle. It arises from the scapula and inserts upon the olecranon process of the ulna. It is the largest of the three heads of the triceps brachii and acts together with the other two heads to extend the forelimb.

Also seen in the photo are the *radial nerve*, *median nerve*, *ulnar nerve*, and the *brachial artery*.

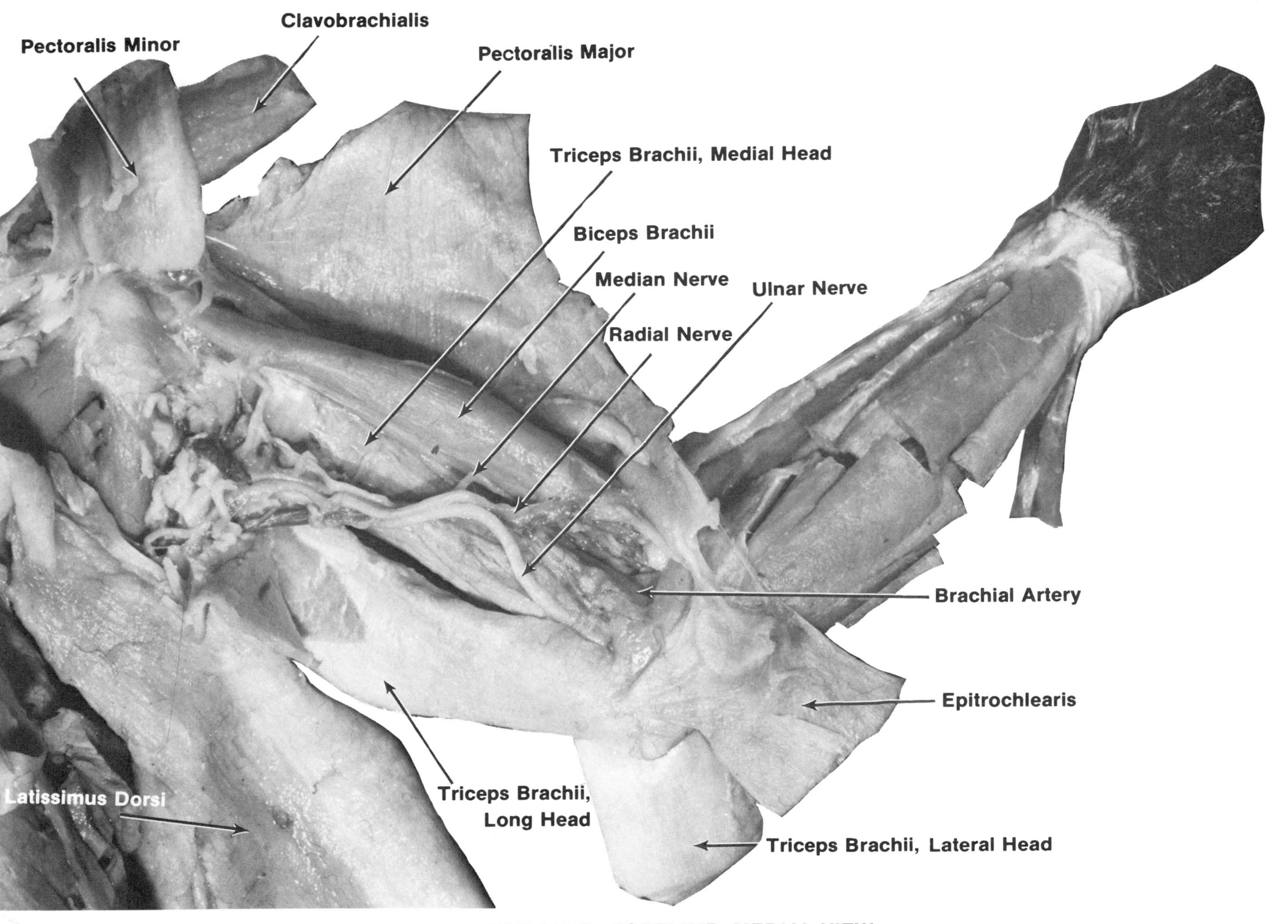


DEEP MUSCLES – FORELIMB, MEDIAL VIEW

# DEEP MUSCLES - LOWER FORELIMB, MEDIAL VIEW

The photo shows some of the superficial muscles above the elbow. These include the:

- pectoantebrachialis
- pectoralis major
- epitrochlearis

The following muscles of the lower forelimb have been transected and reflected to expose the deeper muscles:

- flexor carpi ulnaris
- palmaris longus
- flexor carpi radialis

Observe the muscle that has been exposed.

**Flexor Digitorum Profundus** - This is not a single muscle but an entire group. They appear as five separate muscles or heads, fused with one another. They are attached to the five digits and enable them to move. Three heads originate from the medial epicondyl of the humerus, one from the radius, and one from the ulna. They insert upon the bases of the distal phalanges of the digits. The group acts to flex all of the digits.

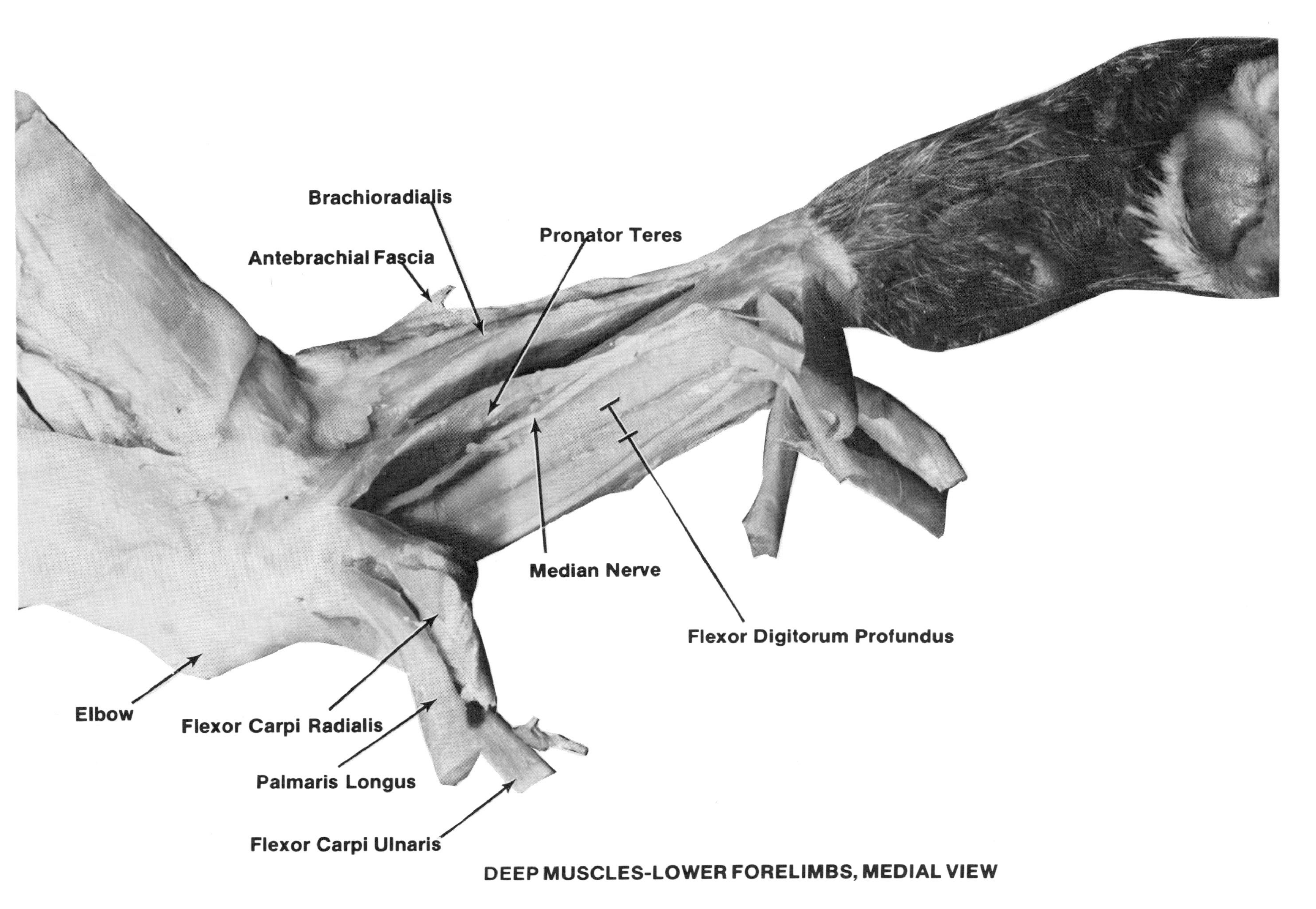


DEEP MUSCLES-LOWER FORELIMBS, MEDIAL VIEW

# DEEP MUSCLES - HIND LIMB, MEDIAL VIEW

Transect the *sartorius* and *gracilis* muscles as in the photo. This will expose the deeper muscles of the hind limb. Parts of some of the muscles have already been seen in a previous photo, page 35, and were described earlier (see page 33). These include the:

adductor femoris
adductor longus
pectineus

They are all adductors of the thigh. In this view much more of each muscle is seen and their actions as adductors become more apparent when we examine the directions of their fibers, their origins, and insertions. They originate from the pubis and insert upon the femur.

The *femoral artery* and *femoral vein* are clearly visible passing down the midline of the limb, between the anterior and posterior portions.

Posterior to the adductor group lies another group of muscles known as the *hamstrings*. The name originated from the practice of butchers who hung hams by the tendons of these muscles. Included are the:

**Semimembranosus** - This muscle is clearly visible here deep to the gracilis. It is a large muscle covering almost half of the posterior thigh. It originates from the ischium and inserts upon the distal end of the femur. It acts to extend the thigh.

**Semitendinosus** - This is the most posterior of the group. It too arises from the ischium and inserts by means of a tendon upon the tibia. It is a flexor of the knee. The relationship between the semimembranosus and semitendinosus muscles is most clearly seen in the photo on page 67.

**Biceps Femoris** - The third muscle of the hamstrings group is located on the lateral surface and cannot be seen in this photo of the medial surface. It was seen in the photo on page 44, and described on page 43.

The hamstrings in man consist of the same muscles, located on the posterior aspect of the thigh. Their actions too are similar.

Another group of muscles occupies the anterior portion of the thigh, the *quadriceps femoris*. It consists of four muscles which are best observed after transecting the sartorius and tensor fascia lata muscles. Included in this group are the:

**Rectus Femoris** - This is a cigar-shaped muscle which arises from the iliac spine and from just above the acetabulum. It lies between the *vastus lateralis* and *vastus medialis*. It acts as a flexor of the thigh as well as an extensor of the leg.

**Vastus Lateralis** - This is a large muscle arising from the lateral and dorsal surfaces of the femur and from the greater trochanter. Part of it lies under the tensor fascia lata muscle. It is the largest of the quadriceps group.

**Vastus Medialis** - This large muscle, the third of the quadriceps group, arises from the dorsal surface of the femur. It lies medial to the rectus femoris.

**Vastus Intermedius** - This last muscle of the quadriceps group, is flat in shape and lies deep to the rectus femoris, between the vastus lateralis and vastus medialis. To observe this muscle lift the rectus femoris. It is not seen in the photo. Its origin is from the ventral surface of the femur and inserts upon the capsule of the knee joint.

The entire group has a common insertion by a large tendon which passes over the patella, the kneecap, to insert on the tuberosity of the tibia. The portion of the tendon between the patella and the tibia is known as the *patellar ligament.*

In combination the quadriceps femoris extends the knee joint. The muscle group is the same as that found in man.

**In the space below make a list of all the deep muscles you have found in medial view of your cat.**

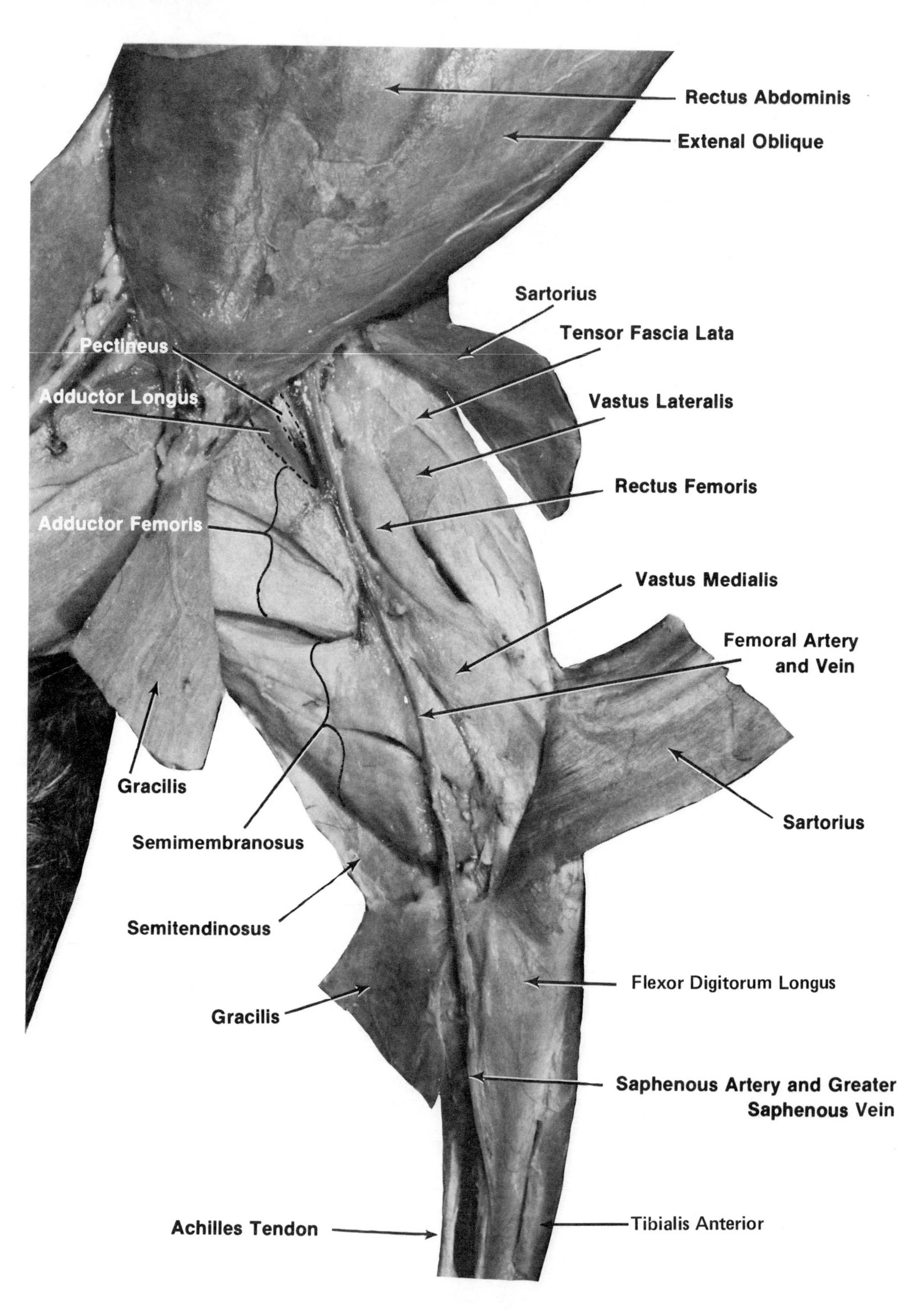


DEEP MUSCLES - HIND LIMB, MEDIAL VIEW

# DEEP MUSCLES
# THORAX AND SHOULDER, DORSAL VIEW

Transect and reflect the following superficial muscles of the thorax and shoulder:

clavotrapezius
acromiotrapezius
spinotrapezius
acromiodeltoid
spinodeltoid

This will expose the dorsal surface of the scapula and the muscles associated with its movement.

**Supraspinatus and Infraspinatus** - The spinous process of the scapula separates two large muscle bundles. The one above the spine, known as the *supraspinatus*, occupies the *supraspinous fossa* of the scapula, while the one below the spine, the *infraspinatus* occupies the *infraspinous fossa.* They lie deep to the acromiotrapezius muscle. They both insert on the greater tubercle of the humerus. The supraspinatus acts to extend the scapula while the infraspinatus rotates the humerus outward.

**Teres Major** - This is a thick muscle which lies posterior to the infraspinatus. It acts to rotate the humerus and draw it posteriorly. Although this muscle was previously examined during the dissection of the ventral surface, it is quite prominent on both surfaces, covering the posterior border of the scapula and much of its dorsal and lateral surfaces. It inserts in common with the latissimus dorsi on the proximal end of the humerus.

**Teres Minor** - This is a much smaller muscle. It originates on the axillarly border of the scapula and inserts on the greater tuberosity of the humerus. It assists the infraspinatus muscle in rotating the humerus laterally. Between the teres major and teres minor lies the origin of the lateral head of the triceps brachii.

Three distinct *rhomboideus* muscles are present. They are clearly seen in the photo. They attach the scapula to the vertebral column. The name is derived from their rhomboidal shape.

**Rhomboideus Major** - This muscle originates from thoracic vertebrae to insert upon the inferior angle of the scapula.

**Rhomboideus Minor** - This portion of the rhomboideus, despite its name, is larger and more extensive than the rhomboideus major. Its name is based upon its relative size in man, where the rhomboideus major is the larger muscle. It originates on the lower cervical and the thoracic vertebrae and inserts on the dorsal border of the scapula.

**Rhomboideus Capitis** - This is the most anterior and lateral of the group. It originates from the occipital area of the skull and inserts on the vertebral border of the scapula.

The combined action of the rhomboid group of muscles is to draw the scapula toward the vertebral column and forward.

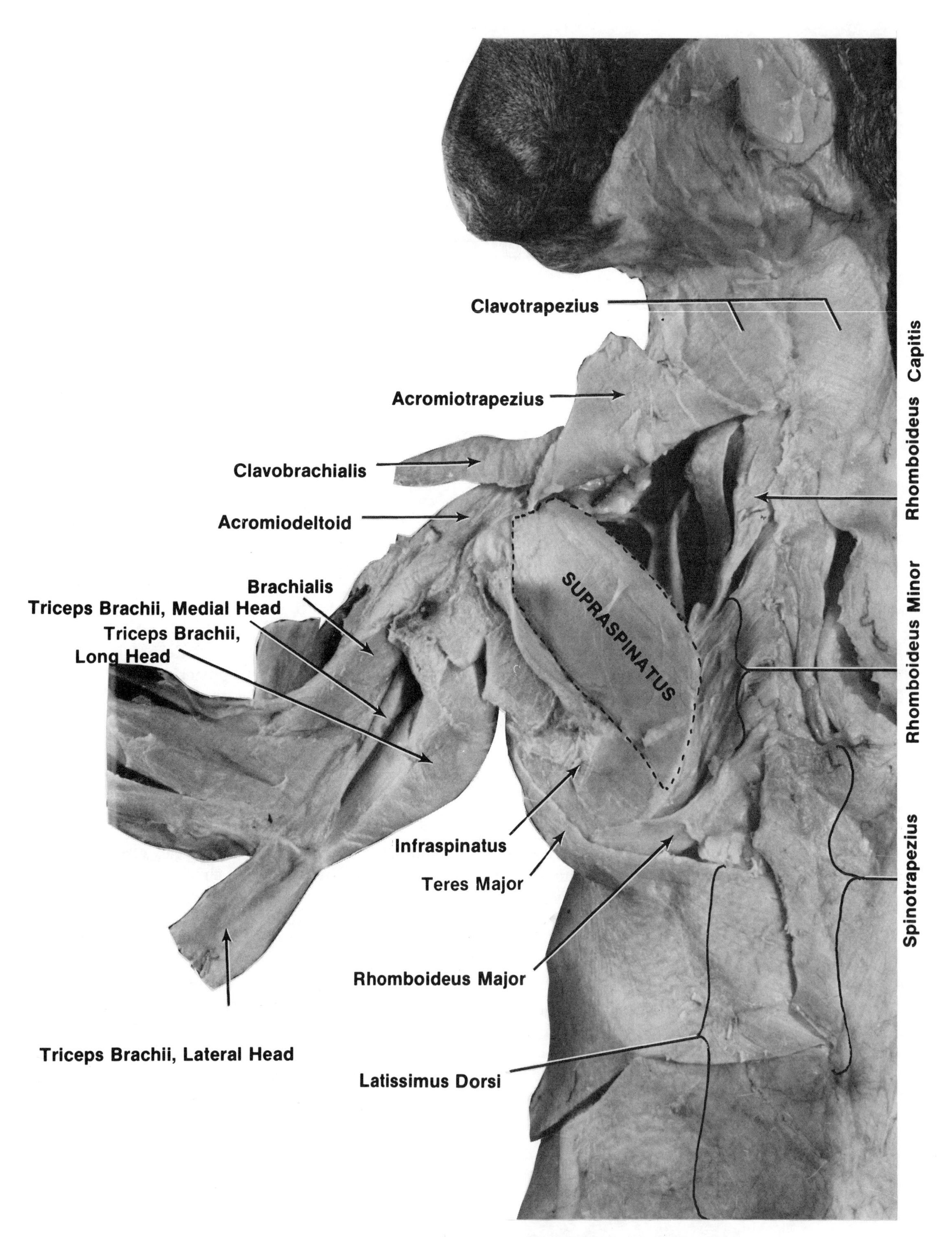


DEEP MUSCLES - THORAX AND SHOULDER, DORSAL VIEW

# DEEP MUSCLES - FORELIMB

## LATERAL VIEW

The next photo is a close-up view of the entire forelimb.
Note that the lateral head of the triceps brachii muscle has been transected. We clearly see the:

- triceps brachii, long head
- triceps brachii, medial head
- teres major
- infraspinatus
- spinodeltoid

Two muscles not previously seen are exposed and clearly visible in the photo.

**Brachialis** - Although a small part of this muscle was seen superficially, the major portion could not be seen until the lateral head of the triceps brachii is transected and reflected. It is now readily visible. It originates upon the humerus and inserts on the proximal end of the ulna. It acts as a flexor of the forelimb. It is much larger and more prominent in the cat that in the human.

**Anconeus** - This small, flat muscle is not found in humans. It may be considered a fourth head of the triceps brachii. It arises from the epicondyl of the humerus and inserts on the olecranon process of the ulna immediately distal to the insertion of the three main heads of the triceps brachii.

The remaining muscles in this area have already been described.

## LOWER FORELIMB, LATERAL VIEW

Transect and reflect the following muscles as in the photo on page 63.

- triceps brachii, lateral head
- extensor digitorum communis
- extensor digitorum lateralis

Several deep muscles may now be observed:

**Extensor Pollicis Brevis** - This muscle is much more extensive in the cat than in the human. Thus, the term "brevis" is misleading. It extends from the distal anterior surface of the ulna and from the lateral surface of the radius. Its fibers converge to form a flat tendon that passes on to the anterio-lateral surface of the wrist. It inserts on the first metacarpal. It acts to extend and abduct the thumb.

**Extensor Indicus Proprius** - This muscle arises caudad to the origin of the extensor carpi ulnaris on the anterio-lateral surface of the ulna. Its fibers extend distally, tight against the ulna. It attaches to the second digit by a white shiny tendon. The muscle acts to extend the second digit and often assists in the extension of the thumb.

**Supinator** - This muscle is not seen in the photo. It is flat and slightly spiral. It arises from the humerus and upper radius and extends down the radius to insert on the ventral surface of the radius. Its diagonal fibers enable the muscle to act as a powerful supinator of the paw.

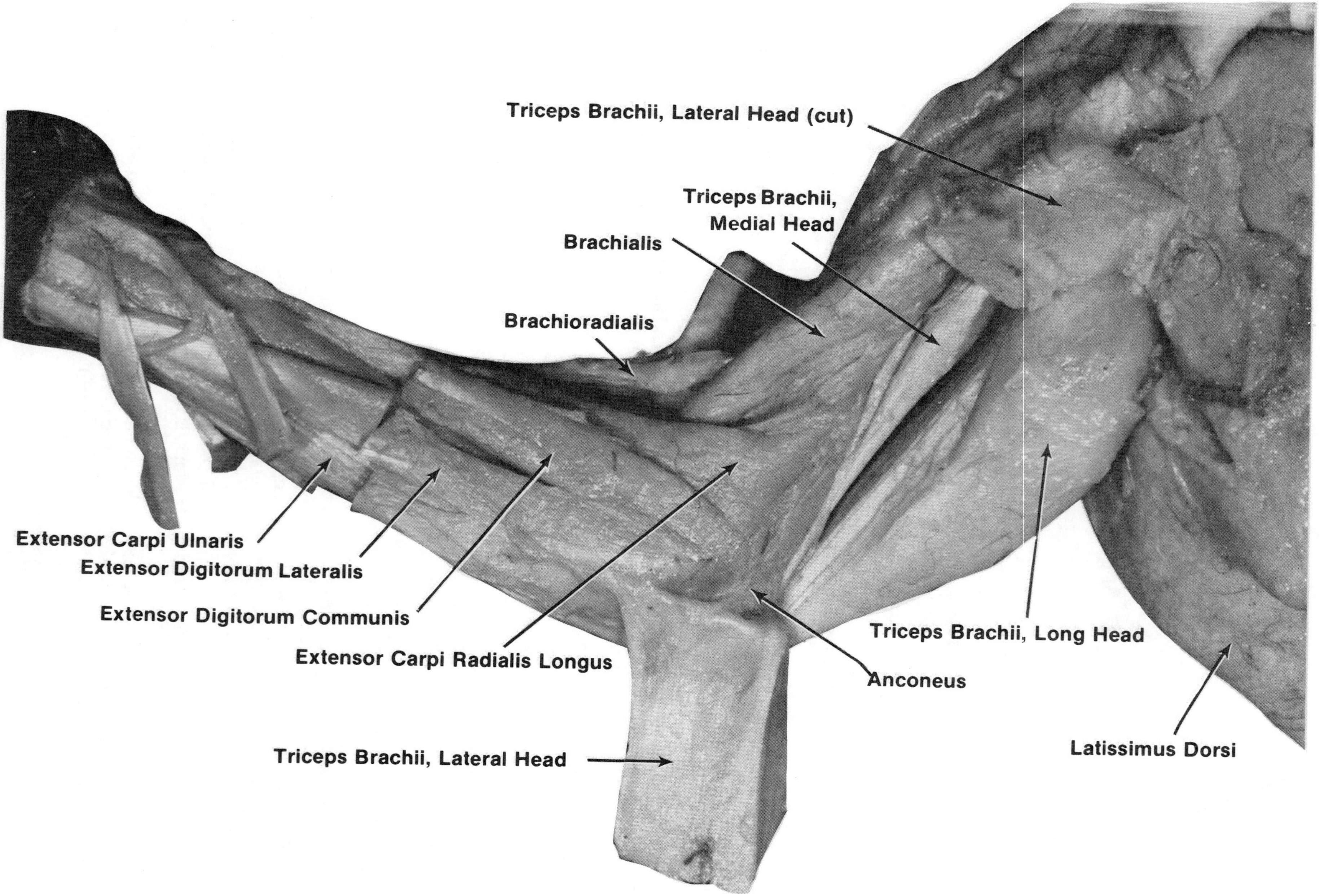

Triceps Brachii, Lateral Head (cut)
Triceps Brachii, Medial Head
Brachialis
Brachioradialis
Extensor Carpi Ulnaris
Extensor Digitorum Lateralis
Extensor Digitorum Communis
Extensor Carpi Radialis Longus
Triceps Brachii, Lateral Head
Triceps Brachii, Long Head
Anconeus
Latissimus Dorsi

DEEP MUSCLES - FORELIMB, LATERAL VIEW

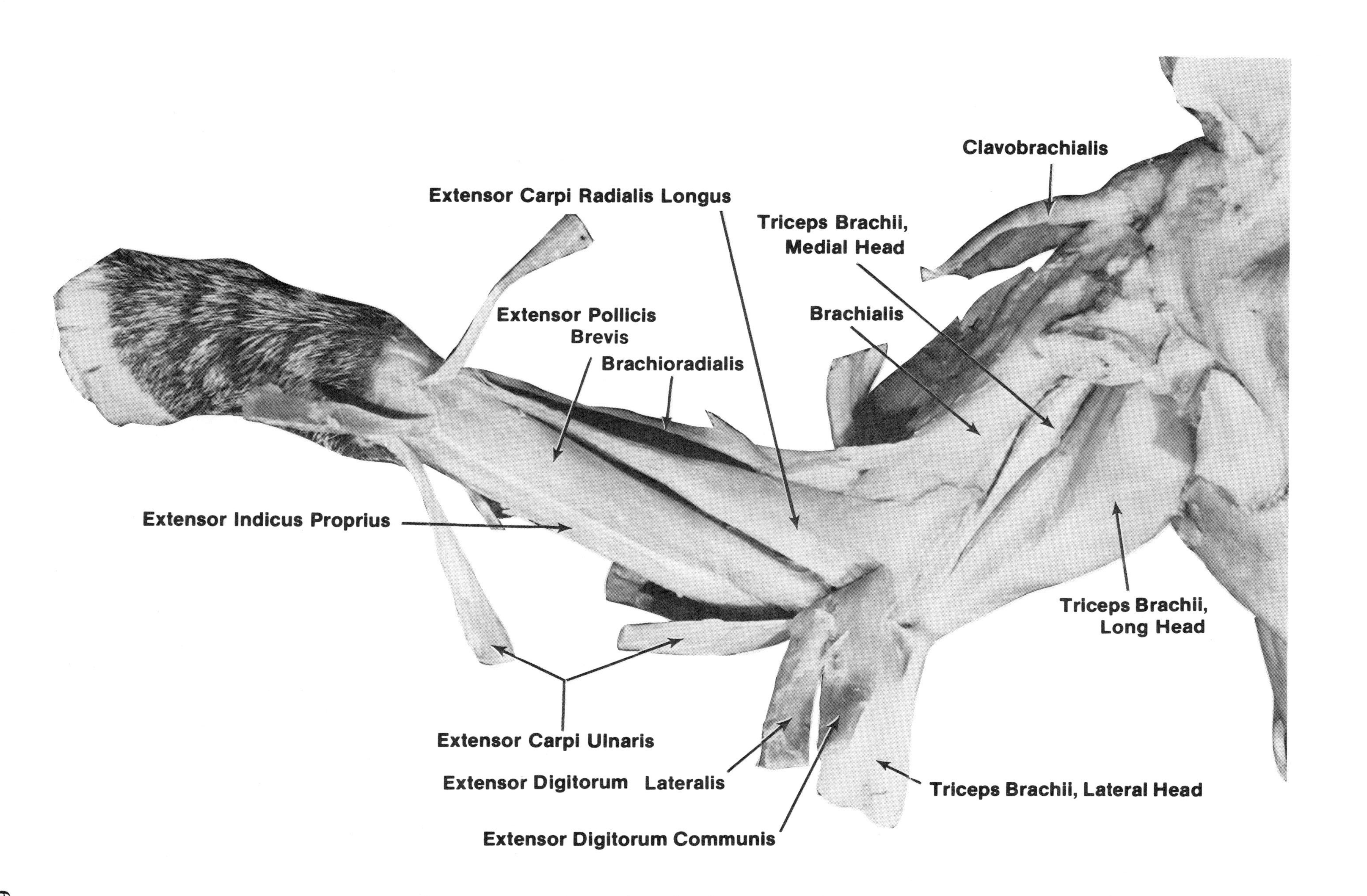


DEEP MUSCLES - LOWER FORELIMB, LATERAL VIEW

# DEEP MUSCLES - LUMBAR AND SACRAL AREAS, DORSAL VIEW

Transect and reflect the following muscles:
spinotrapezius
latissimus dorsi

In addition, the very prominent *lumbodorsal fascia,* the white aponeurosis covering most of the lumbodorsal surface is cut to reveal the deeper muscles below.

**Extensor Dorsi Communis** - This large muscle mass on each side of the vertebral column extends from the sacrum and ilium to the skull. As its name indicates, it is an extensor of the spine. It also draws the ribs posteriorly and bends the neck and spinal column to one side. It is comparable to the *sacrospinalis (erector spinae)* of humans.

It is divided longitudinally into three columns of muscle tissue:

**Iliocostalis** - This is the most lateral of the three. In the thoracic region it inserts upon the ribs and in the lumbar region upon the transverse processes of the vertebrae. It arises from the ilium, the lumbar vertebrae, and the ribs.

**Longissimus** - This bundle is medial to the iliocostalis. It extends to the skull. Different names for the various segments are used according to their location. These are the *longissimus dorsi, cervicis,* and *capitis.*

**Spinalis** (Spinalis Dorsi) - This is the most medial of the three muscle columns. It consists of diagonal fibers which insert on the spinous processes of the upper lumbar, thoracic, and cervical vertebrae.

Three other dorsal muscles are:

**Splenius** - This muscle is located on the back of the neck. It lies deep to the clavotrapezius and rhomboideus muscles. It arises from the first two thoracic vertebrae and from the dorsal midline of the neck to insert upon the occipital bone of the skull. It is an extensor of the head and flexes it laterally.

**Multifidus** - This muscle extends the entire lenght of the **vertebral column,** from the sacrum to the skull, filling the grooves adjacent to the spinous processes of the vertebrae. It may be observed in the lumbar region medial to the longissimus dorsi. It acts to erect and rotate the spine.

**Semispinalis** - This muscle lies deep to the splenius. It has two sections, the *semispinalis cervicis* and *capitis.* They arise from the vertebrae between the anterior portion of the spinalis. These muscles extend forward to the neck and head.

The deep muscles of the back are all concerned with the extension and lateral flexion of the back, neck, and head. They are called *epaxial,* and are innervated by the smaller dorsal rami of the spinal nerves. The other skeletal muscles of the cat are *hypaxial.*

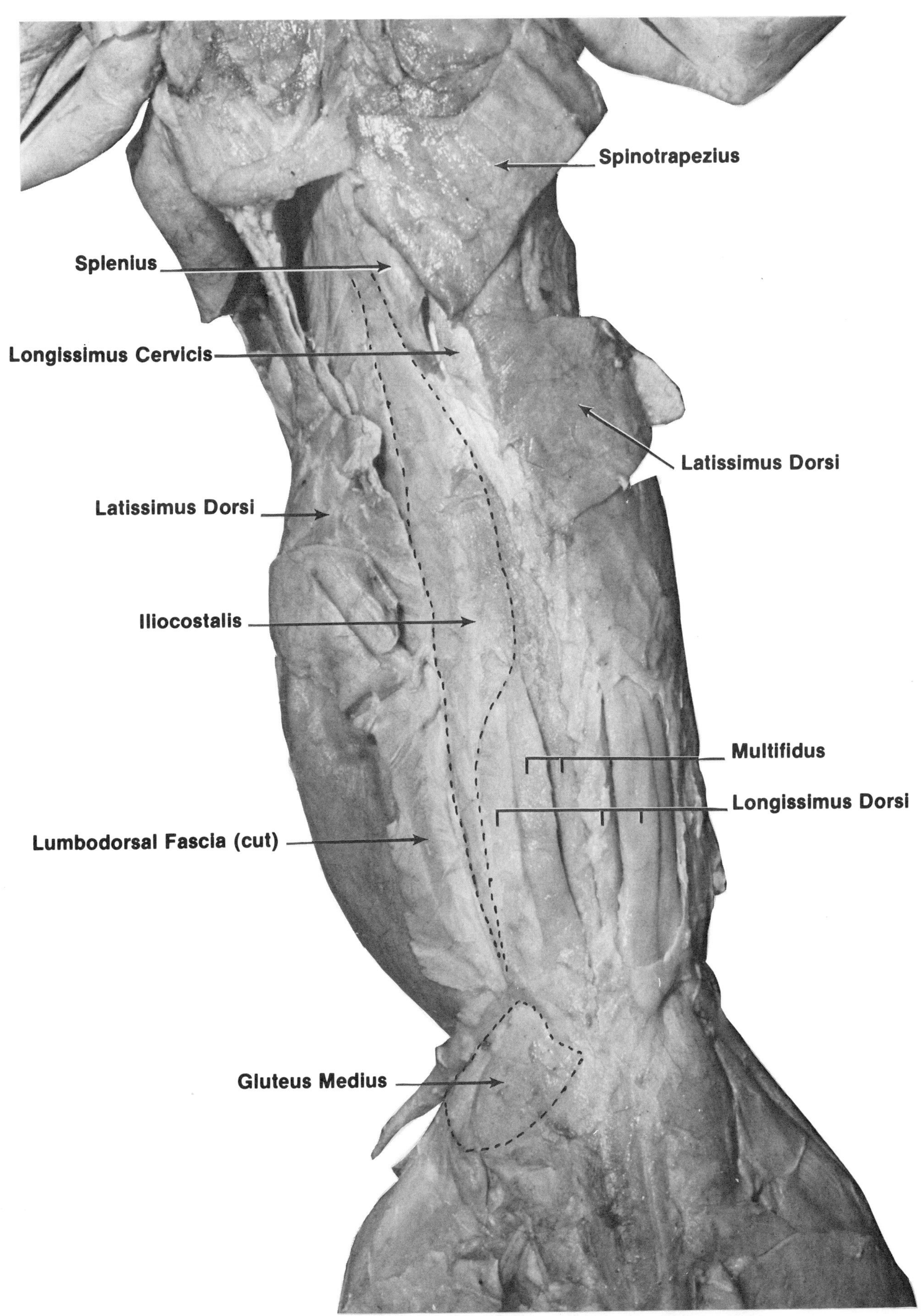


DEEP MUSCLES - LUMBAR AND SACRAL AREAS, DORSAL VIEW

# DEEP MUSCLES - THIGH AND HIND LIMB, LATERAL VIEW

In this dissection the large *biceps femoris* muscle of the thigh is transected and reflected.

Many muscles previously seen only superficially are now observed in different view. These include the:

semimembranosus
semitendinosus
vastus lateralis
adductor femoris

These muscles have already been described in previous sections.

The relationship between the semimembranosus and the semitendinosus muscles is clearly seen in this photo. The space between them is exaggerated.

The most prominent new structure observed is the *sciatic nerve*. It runs diagonally across the center of the photo. It is one of the thickest nerves of both the cat and man. As seen in the photo it originates in the sacrum, passes through the thigh to innervate the gastrocnemius muscle. Small brances innervate the semimembranosus, the tensor fascia lata, as well as other muscles of the thigh.

**Tenuissimus** - This slender band of muscle runs parallel to the sciatic nerve, somewhat posteriorly. It originates from the second caudal vertebra and inserts on the tibia in common with the biceps femoris. It assists the biceps femoris in abduction of the thigh and flexion of the shank. It has no homolog in man. Only a small portion of this muscle is seen in the photo.

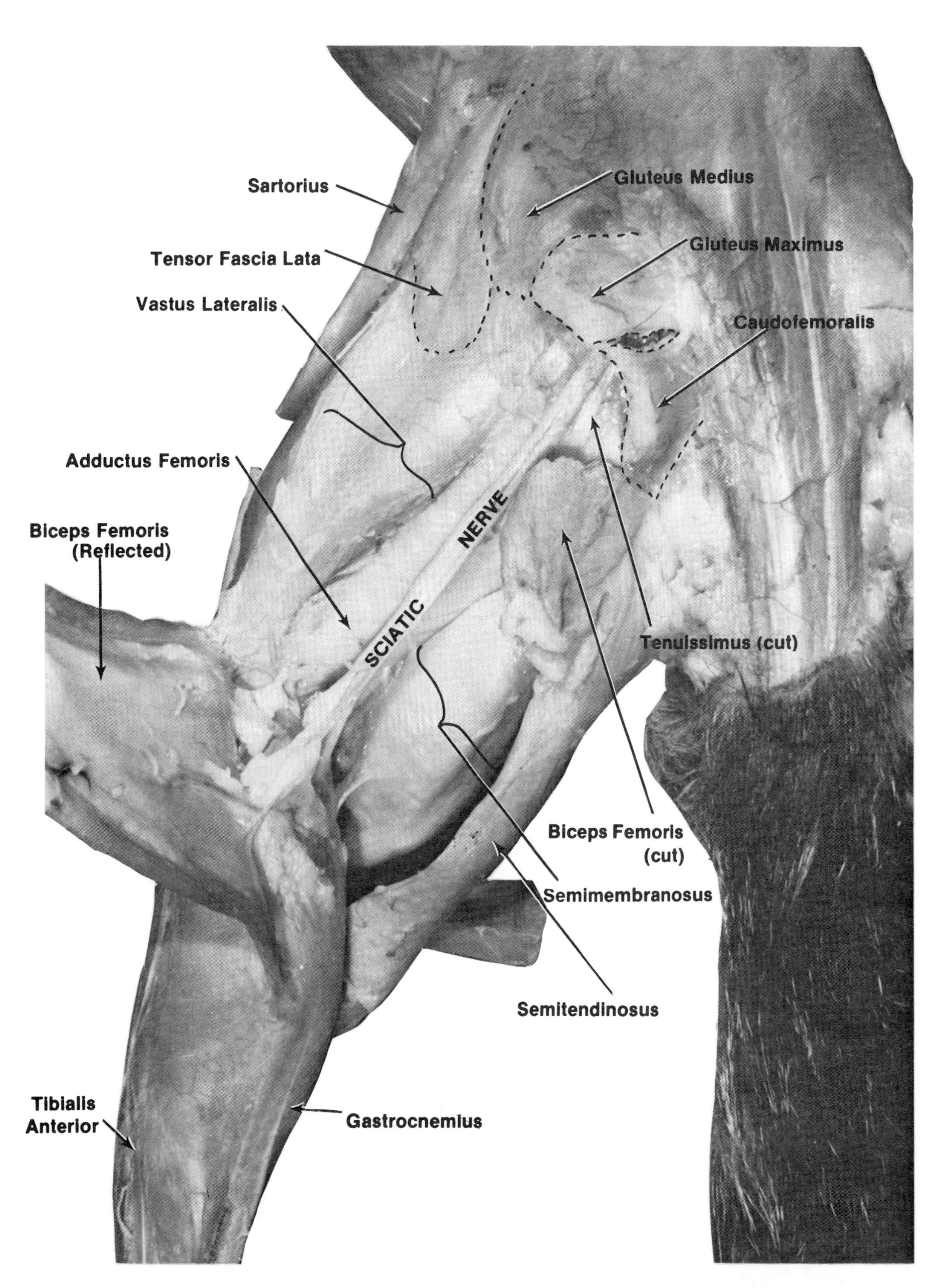


DEEP MUSCLES - THIGH AND HIND LIMB, LATERAL VIEW

# HUMAN MUSCULAR SYSTEM

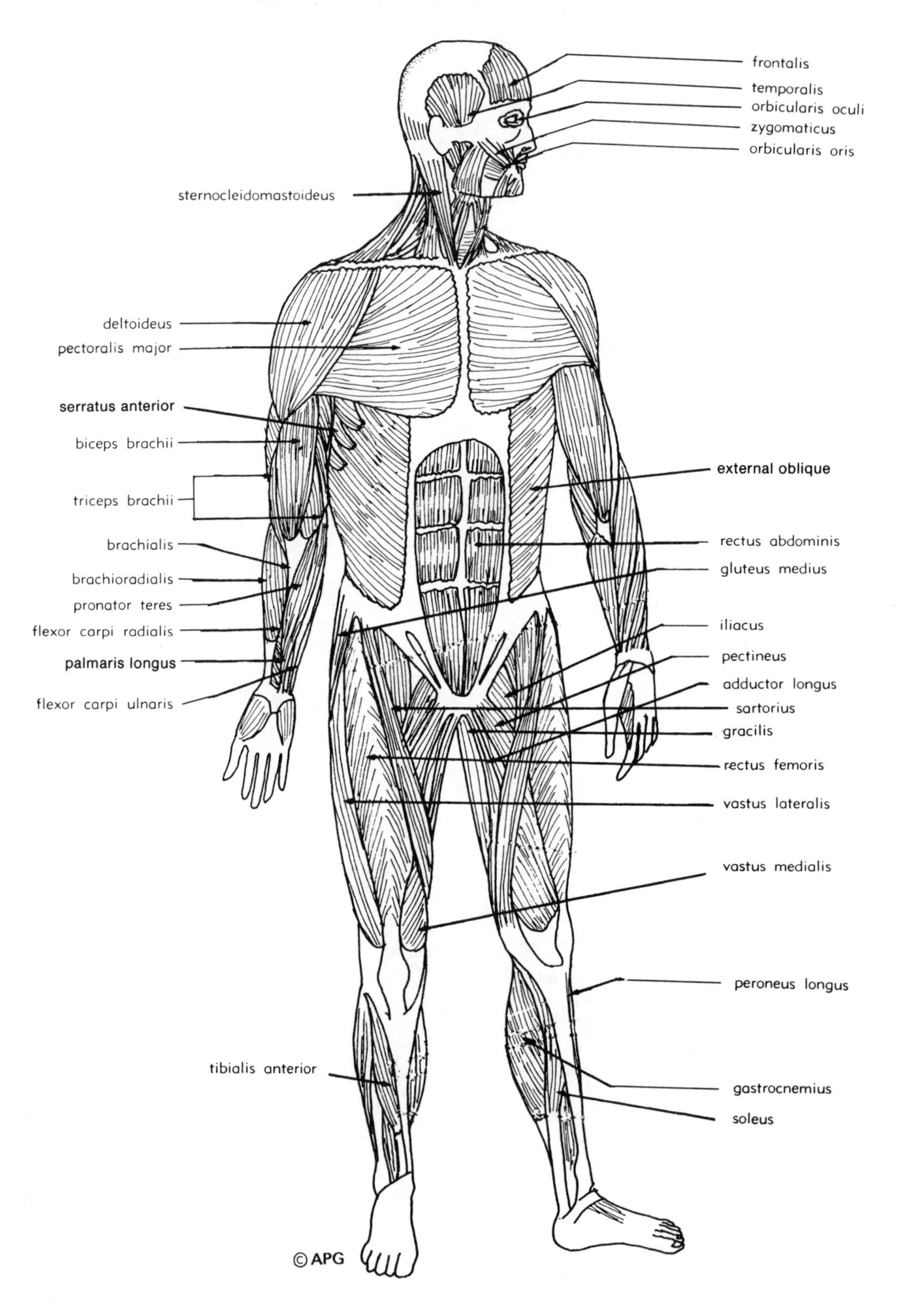


**ANTERIOR VIEW**

# HUMAN MUSCULAR SYSTEM

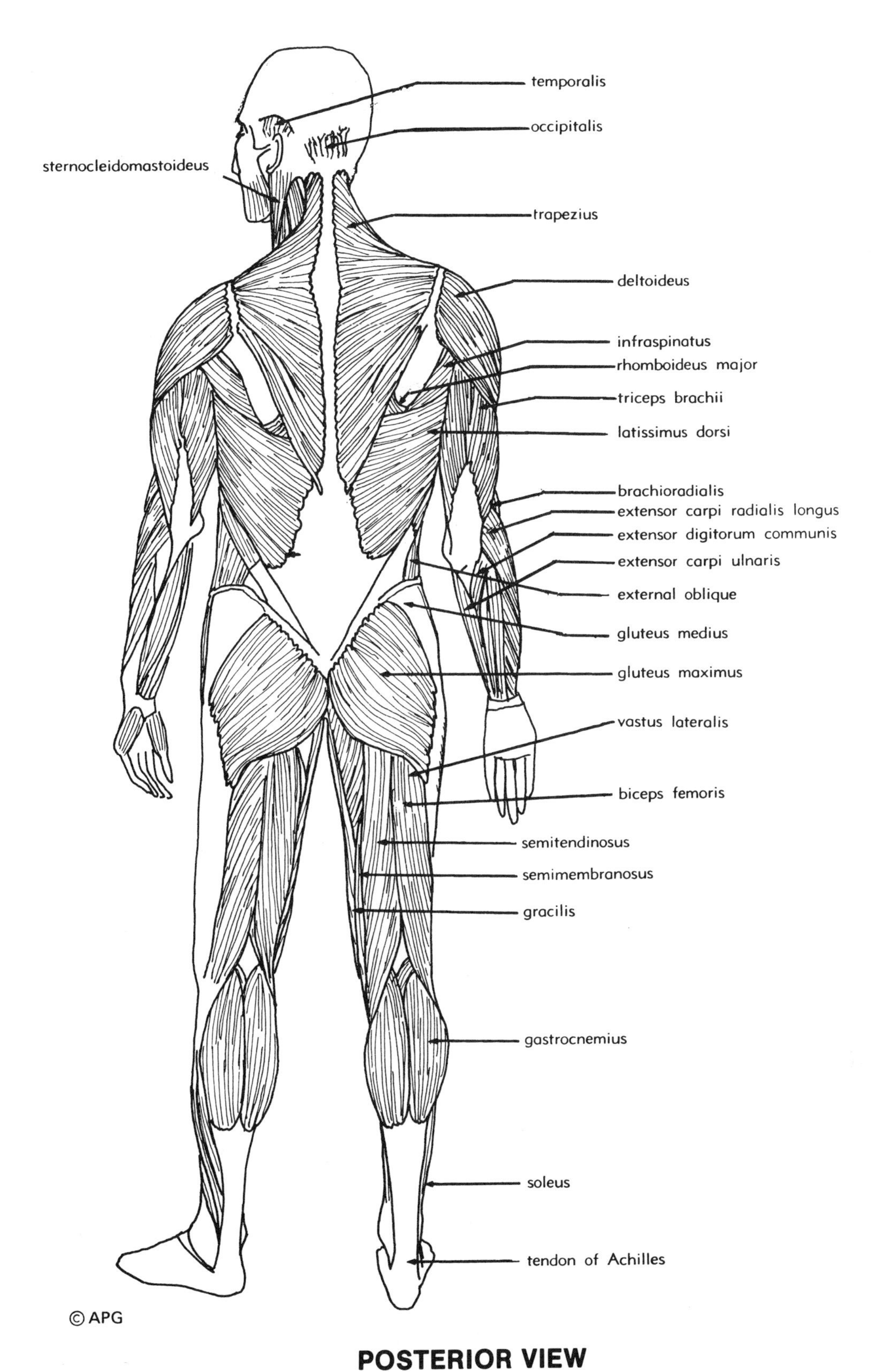




**POSTERIOR VIEW**

Name .................................. Section ....... Date ................

# SELF - QUIZ I
# MUSCULAR SYSTEM

1. The skin is connected to the muscle layers below by a fibrous connective tissue known as ....
2. The bone to be moved by a skeletal muscle serves as the, a) origin, b) insertion, c)action, d)rotation, for that muscle.
3. Name the muscle that acts as the antagonist of the biceps brachii.
4. The ventral and lateral portions of the abdominal wall are composed of three layers of muscle. Name these in the proper order beginning with the outermost one.
5. Name the major superficial muscles of the shoulder and dorsal thoracic areas.
6. Name the tough connective tissue that ties the gastrocnemius muscle to the calcaneous bone.
7. Name the muscles that raise the ribs; that lower the ribs.
8. Name the large, broad sheet of muscle that originates along the thoracic and lumbar vertebrae and inserts upon the humerus.
9. The sartorius and gracilis are superficial muscles located at the, a)head, b)chest, c)arm, d)leg.
10. Define each of the terms listed below.

ANSWERS

1. ______________________________
2. ______________________________
3. ______________________________
4. ______________________________
5. ______________________________
6. ______________________________
7. ______________________________
8. ______________________________
9. ______________________________
10.
    a. aponeurosis ______________________
    b. pronation ______________________
    c. flexion ______________________
    d. circumduction ______________________
    e. linea alba ______________________
    f. transect ______________________
    g. orbicularis oris ______________________
    h. masseter ______________________
    i. hamstrings ______________________
    j. synergistic ______________________

Label all of the features indicated on the photograph

# THE SALIVARY GLANDS

The salivary glands were observed when you dissected the musculature of the head. If the glands were not injured they may now be studied in detail. If, however, they were destroyed, begin as follows.

Carefully remove the skin on the side of the head over-lying the cheek, throat, and side of the neck below the pinna of the ear. The *platysma* and other superficial facial muscles associated with the skin are removed. Work carefully, especially in the cheek region, in order not to injure the glands or their narrow ducts.

**Parotid Gland** - This is the largest of the salivary glands. It lies ventral to the pinna and is recognized by its triangular shape and lobular texture. The *parotid duct* can be seen emerging from the anterior edge of the gland, then crossing the large *masseter muscle* of the cheek and continuing to the oral cavity, where it opens opposite the last upper premolar. You can find the opening by looking inside the cheek and tugging lightly on the duct with forceps.

Note: Do not confuse branches of the facial nerve leading to facial muscles with the parotid duct. They too cross the masseter muscle from beneath the edges of the parotid gland, one dorsal the other ventral to the parotid duct. The parotid duct is generally thicker than the nerve branches. The relationship between the nerves and the parotid duct can clearly be seen in the photo.

**Submaxillarly Gland** - This gland lies immediately ventral to the parotid, posterior to the angle of the jaw. It is smaller than the parotid and oval in shape. Do not confuse it with the still smaller lymph glands in this area. Distinguish one from the other by observing their surface textures. Lymph nodes are smooth, salivary glands are lobulated. The *submaxillary duct* emerges from the front of the gland and passes forward to digastric and mylohyoid muscles. Cut these muscles to observe the duct into the floor of the mouth just behind the lower incisors. The duct opens on a pair of papillae at the base of the tongue.

**Sublingual Gland** - This is a small oblong gland located anterior to the submaxillary gland. The duct of this gland and of the submaxillary accompany one another and have a common opening into the mouth. Lymph glands overlap parts of this gland.

Two other salivarly glands are found in the cat.

**Molar Gland** - This gland located near the corner of the mouth is small and diffuse. It is found between the masseter and the mandible. It opens into the mouth by many small ducts.

**Infraorbital Gland** - This gland is located on the floor of the orbit of the eye.

Note the cat's teeth. They are all sharp and pointed. The dentition of the cat in relation to its diet as well as comparisons to other mammals were made earlier in the section on the skeleton.

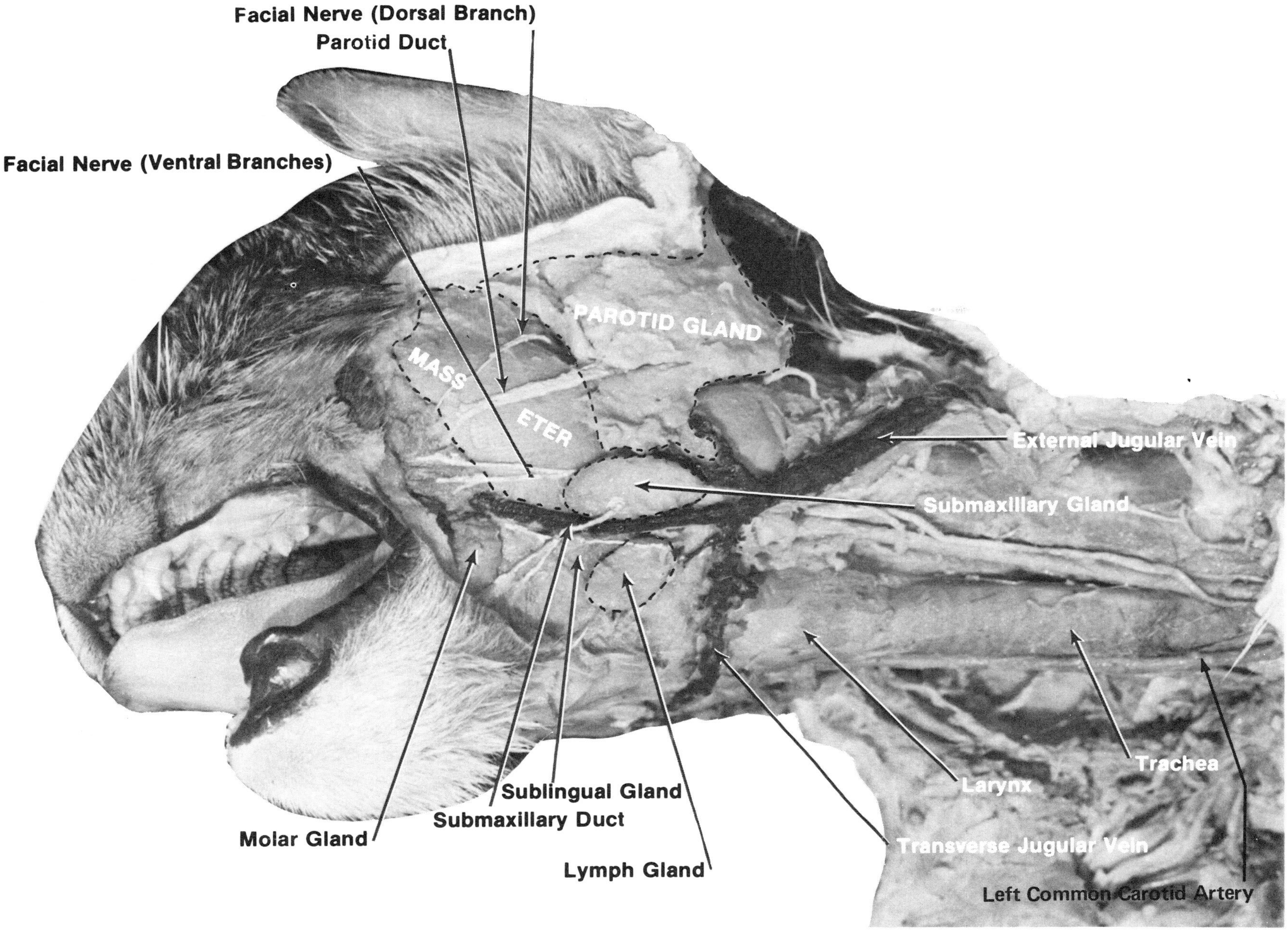

THE SALIVARY GLANDS

# THE ORAL CAVITY

With your scissors cut through the muscles at both corners of the mouth in a posterior direction. Then use a bone clipper or bone saw to cut through the angles of the jaw. This will expose the following:

**Vestibule** - This is the region between the cheeks and the teeth.

**Tongue** - The large muscular tongue lies upon the floor of the mouth. It is connected to the floor by the vertical *lingual frenulum.* The surface of the tongue is covered by variously shaped projections known as *papillae.* The most numerous of these are the pointed *filiform papillae.* The anterior ones bear spiny projections with which the cat grooms its fur and rasps flesh from bones. Use a hand lens to observe them more closely. Microscopic *taste buds* are found at the sides and base of the papillae.

**Palate** - This structure forms the roof of the mouth. It is a partition which separates the oral and nasal cavities.

The anterior portion is supported by bone and is known as the *hard palate.* A series of eight transverse ridges, the *palatine rugae,* containing numerous papillae, cross the roof of the mouth.

The *soft palate* continues beyond the hard palate posteriorly. It is a muscular structure which divides the *oral pharynx* ventrally from the *nasal pharynx* dorsally. In man, there is a finger-like process, the *uvula,* which hangs down from its center posteriorly. It is absent in the cat. The *palatine tonsils* are small rounded masses of lymphatic tissue located in the lateral walls of the oral pharynx below the soft palate near the base of the tongue. Only a small portion of each is visible, the remainder is embedded in the mucous membrane.

The large opening from the oral to the nasal pharynx is the *isthmus of fauces.* It is located upon the surface of the soft palate, dorsal to the glottis and the top of the esophagus. The *eustachian tubes* pass from the dorsal nasal pharynx to the middle ear. The *internal nares* open to the anterior end of the nasal pharynx. They are continuous with the *external nares,* or nostrils. Slit the soft palate longitudinally to observe these structures.

**Epiglottis** - This cone-shaped flap of cartilage is located at the top of the *larynx* (voice box) near the base of the tongue. It protects the *glottis,* the slit-like opening to the trachea. During swallowing and eating the epiglottis prevents food from entering the trachea.

**Trachea** - This tube is commonly called the windpipe. It is topped by the epiglottis and larynx. It is kept open by rings of cartilage which extend around the trachea at intervals. They are incomplete dorsally. The trachea branches to form two *bronchi* which enter the lungs.

**Esophagus** - This muscular tube, located dorsal to the trachea is also known as the gullet. Unlike the trachea, however,it is collapsed. Food is pushed forward in the esophagus by the rhythmic contractions of its walls, a process known as *peristalsis.*

In the photo, two wooden probes are prominently visible. The one on the left has penetrated the glottis and has passed down the trachea. Note the epiglottis. A small section of trachea was removed to show the continuity of the wooden prove within the trachea. The probe on the right has entered the esophagus. As you can see, the esophagus lies dorsal to the trachea. A small section of esophagus has similarly been removed to show the continuity of the wooden probe within the esophagus.

**Compare the oral cavity of the cat to that of the human as diagrammed below.**

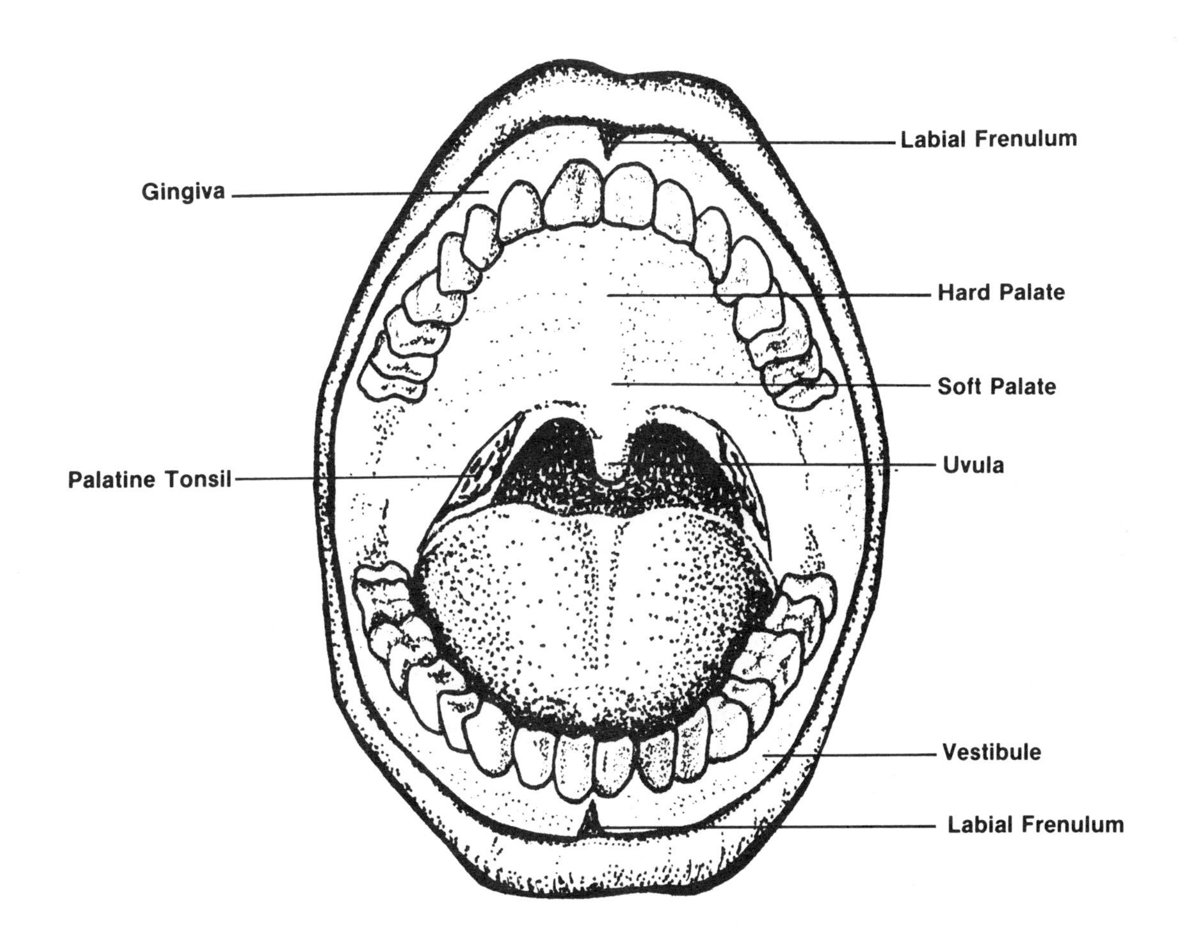


THE HUMAN ORAL CAVITY

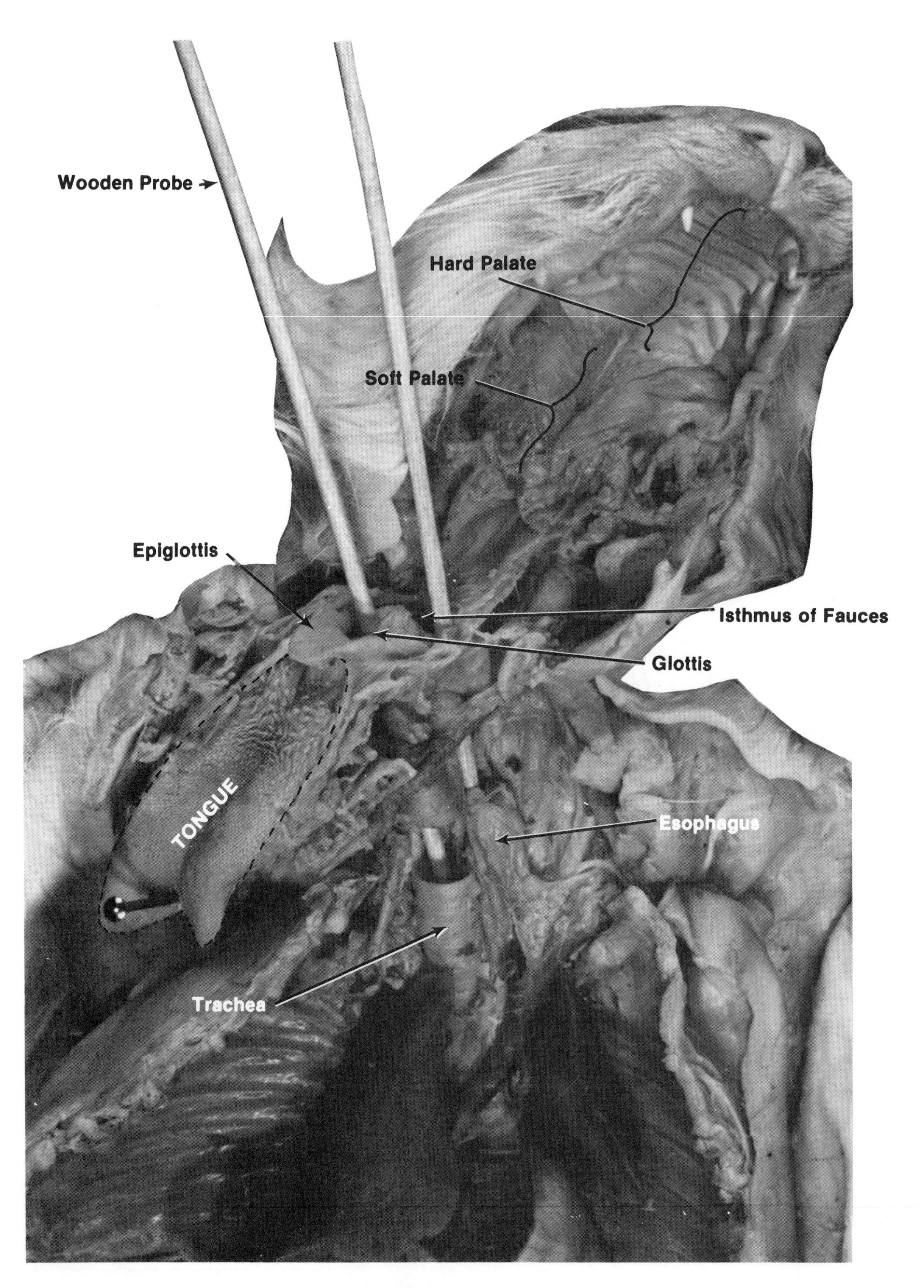


THE ORAL CAVITY

Name ................................ Section ....... Date ................

# SELF - QUIZ II
# THE ORAL CAVITY

1. Name the divisions of the pharynx.
2. Name seven passageways that penetrate the pharynx.
3. Name four different types of papillae found on the surface of the tongue. How do they differ?
4. Name the bones that comprise the hard palate.
5. Besides the palatine tonsils, what other tonsils do you possess?
6. The secretions of the salivary glands begin the digestion of, a) proteins, b) sugars, c)starches, d)lipids.
7. The esophagus is located, a) to the right of the trachea, b) to the left of the trachea, c) dorsal to the trachea, d) ventral to the trachea.
8. A major blood vessel is formed just posterior to the submaxillary gland as a result of the union of smaller vessels. Name the major blood vessel formed.
9. Name the cap of cartilage which prevents food from entering the esophagus while swallowing
10. Define each of the terms listed below.

ANSWERS

1. ______
2. ______
3. ______
4. ______
5. ______
6. ______
7. ______
8. ______
9. ______
10. a. dentition ______
    b. eustachian tube ______
    c. pharynx ______
    d. larynx ______
    e. naso-pharynx ______
    f. frenulum linguae ______
    g. deciduous teeth ______
    h. glottis ______
    i. Stensen's duct ______
    j. Wharton's duct ______

Label all of the structures indicated on the photograph.

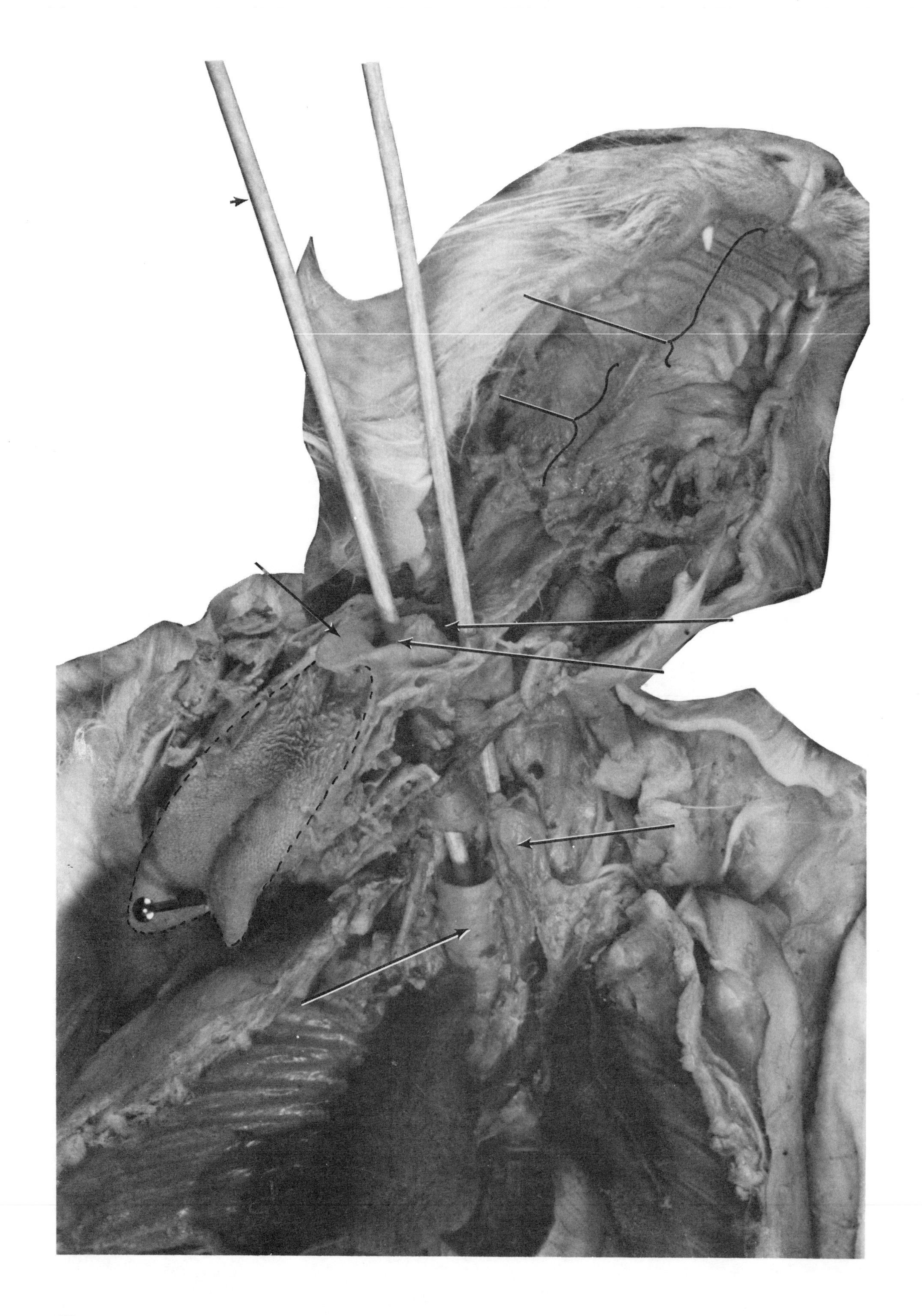

# THE ABDOMINAL CAVITY

The muscular *diaphragm* separates the upper from the lower ventral body cavity. The upper is the *thoracic*, the lower is the *abdominal cavity*. We shall study the abdominal area first and later consider the thorax in relation to the study of the heart and circulatory system.

With your fingertips locate the lower edges of the ribs. Your fingertips will be tracing an arc, an inverted letter "V".

Use your scalpel and cut the musculature along the line you have traced with your fingertips. After the muscle layers have been cut you will find a fine membrane, the *peritoneum*, which lines the inside of the abdominal cavity. The portion of this serous membrane that you see is the *parietal peritoneum*, the *visceral peritoneum* covers the abdominal viscera. Cut through the peritoneum along the mid-ventral line from the xiphoid process to the pubic symphasis. Then cut laterally toward the hind legs. Fold back the abdominal wall to expose the entire abdominal cavity.

Some specimens may contain excess preservative fluid, coagulated blood, or dye that has escaped from the blood vessels. In these cases it is first necessary to wash out the abdominal cavity. Hold the cat under a moderate water flow in the sink and rinse gently. Use paper towels to soak up excess water. Your view should now correspond to that in the photo. Identify the following structures:

**Diaphragm** - This dome-shaped muscular wall separates the thoracic from the abdominal cavity. It is also the most important muscle for *breathing*, permitting inhalation and exhalation. Three major vessels pass through the diaphragm between the thorax and the abdomen. These are the *aorta*, *posterior vena cava*, and the *esophagus*.

**Liver** - This dark brown organ dominates the upper abdomen. The *falciform ligament*, a ventral peritoneal membrane attaches the liver to the diaphragm and to the ventral body wall. The *coronary ligament* attaches the dorsal portion of the liver to the central tendon of the diaphragm.

The falciform ligament lies in a cleft of the liver which divides it into right and left halves, Five lobes can be differentiated: *right lateral*, *right medial*, *left lateral*, *left medial*, and the *caudate lobe*.

**Gall Blader** - This sac-like structure stores bile and releases it into the duodenum. It is located in a depression on the dorsal surface of the right medial lobe. Bile is transported by the *cystic duct* from the gall blader. It is joined by the *hepatic duct* from the liver to form the *common bile duct* which enters the duodenum.

In the photo we observe the abdominal cavity as it appears when we first begin the dissection. Almost all of the abdominal organs below the liver are covered by a double-layered sheet of peritoneum containing many fat deposits. This apron-like membrane is known as the *greater omentum*. A smaller, *lesser omentum* can be seen in the photo on page 83, enclosing the ducts of the liver and gall bladder.

In the second photo the greater omentum has been removed. This reveals abdominal organs which could not be observed earlier. They include the:

**Stomach** - This muscular pouch lies on the left side in the upper abdomen. It is the continuation of the esophagus. Find the esophagus and locate where it pierces the diaphragm to join the stomach. This is the *cardiac* end of the stomach. The *fundus* is the dialated anterior portion, the *body* is the main portion, while the *pyloric* region is the most posterior. This end joins the duodenum.

Open the stomach with your scissors by cutting along the *greater curvature* of the stomach, on the left side. Wash out the contents of the stomach. Note the *cardiac sphincter* which controls the entrance of food into the stomach from the esophagus. The *pyloric sphincter* at the posterior end regulates the release of partially digested food (chyme) into the duodenum. Look along the inner walls of the stomach and note the *rugae*, or folds which help to churn and mix the food with the digestive juices.

**Small intestine** - The first portion of the small intestine is the *duodenum*. It is a continuation of the pyloric end of the stomach. It is a short "U" shaped tube, approximately four inches long. The common bile duct and the pancreatic duct open into the duodenum. The second section of the small intestine is the *jejunum*, which makes up about half the length of this organ. The *ileum* is the final section. Open the jejunum or ileum, wash its contents and touch the inner surface with your fingertips. The velvety texture felt is due to the presence of numerous *villi* along the inner walls. Use a hand lens or a low power dissection microscope to observe them more clearly.

The coils of the small intestine are held in place by a fine peritoneal membrane, the *mesentery*. It is clearly seen in the photo on page 85. Note its shiny thin appearance. It is interlaced with narrow blood vessels, lymphatic vessels, adipose tissue, and lymph nodes. Some of the tiny blood vessels form the beginnings of the portal system, transporting digested food from the intestine to the liver. Cut through the mesentery to unravel the small intestine. Measure its length. How does it compare to the relative length of man's intestine (about twenty feet)?

**Large Intestine** - The end of the ileum projects into the *caecum*, the first segment of the large intestine, about a half inch above its origin. At the junction, the ileocecal sphincter valve will be found. Make a cut here and observe the valve. The *appendix* which extends from the cecum of man is missing in the cat. The entire large intestine consists of the *caecum* and the:

*ascending colon*
*transverse colon*
*descending colon*
*rectum*
*anus*

Locate and identify these segments of the large intestine.

**Pancreas** - The pancreas is a flat elongated gland which lies between the duodenum and the spleen. It is lobular in appearance similar to the parotid salivary gland, but darker in color. Its duct, the *pancreatic duct*, enters the duodenum together with the common bile duct. An accessory pancreatic duct is absent in some specimens. It enters the duodenum at a point posterior to that of the common bile duct.

**Spleen** - This dark-colored organ lies to the left of the stomach, along its greater curvature. It is tied to the stomach by a portion of the greater omentum known as the *gastrosplenic ligament*. It is a part of the circulatory system. It is relatively larger in the cat than in man.

**Kidney** - This reddish, brown bean-shaped organ lies embedded retroperitoneally, namely, behind the parietal peritoneum, in the dorsal body wall, one on each side. In the cat the right kidney lies slightly anterior to the left. In humans the positions are reversed. The *adrenal gland* is located near the anterior end of each kidney, but is separated from it and lies slightly mediad of the kidney. In humans the adrenal gland forms a "cap" upon the kidney.

**Urinary Bladder** - This oval-shaped organ lies at the ventro-posterior end of the abdominal pelvic region. Its size is dependent upon the amount of urine it holds. It is suspended ventrally and laterally by suspensory ligaments. The ventral ligament, the *urachus*, extends from the bladder to the navel and is a remnant of the embryonic allantois. The lateral ligament normally contains fat deposits.

**Reproductive Structures** - Most of the female reproductive structures and some of the male's are located in the abdominal cavity. In the photo, page 84, we see the *ovary, oviduct, horns of the uterus*, and *uterus*.

The urogenital system and its associated structures will be studied and more fully discussed in a later chapter.

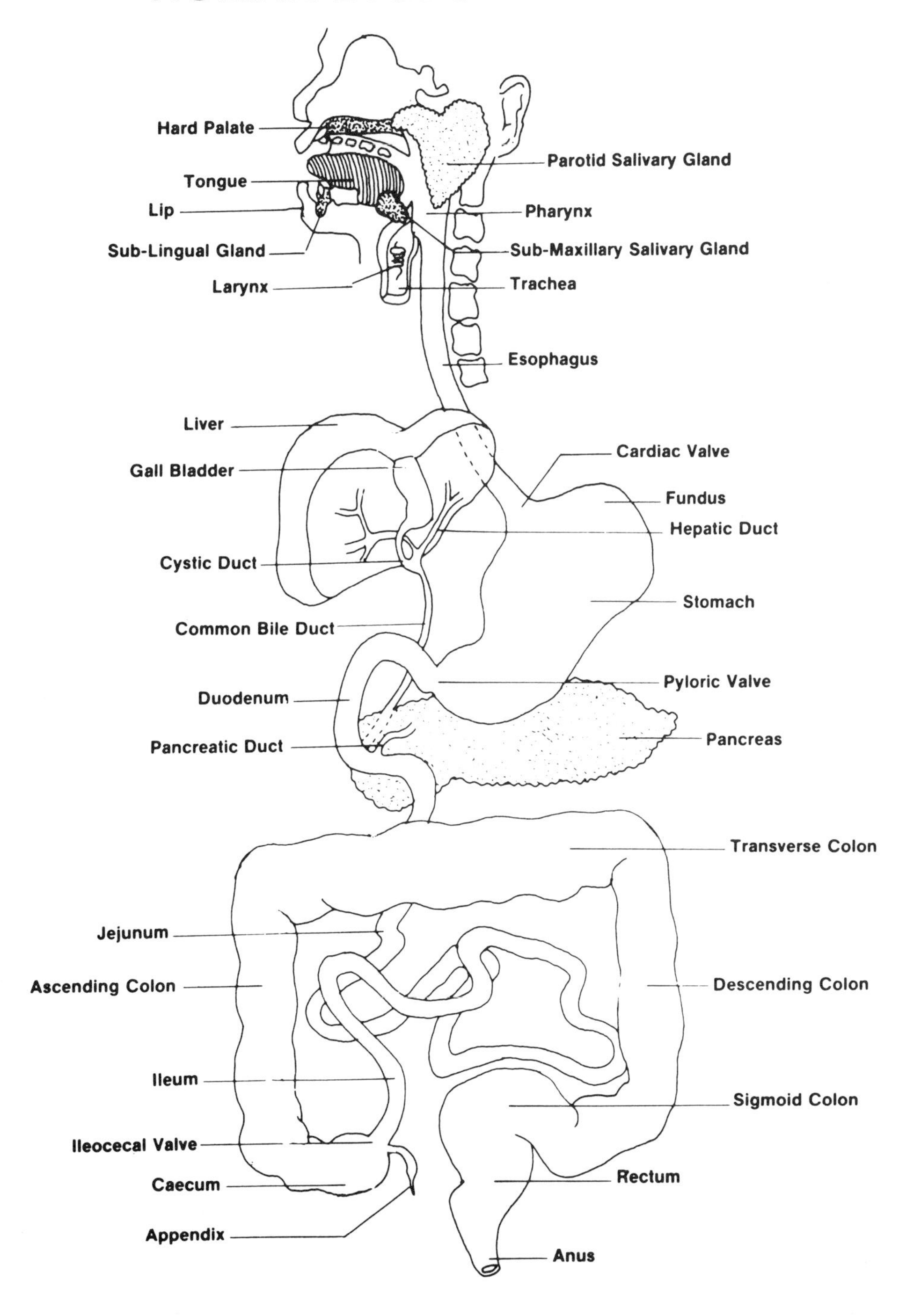

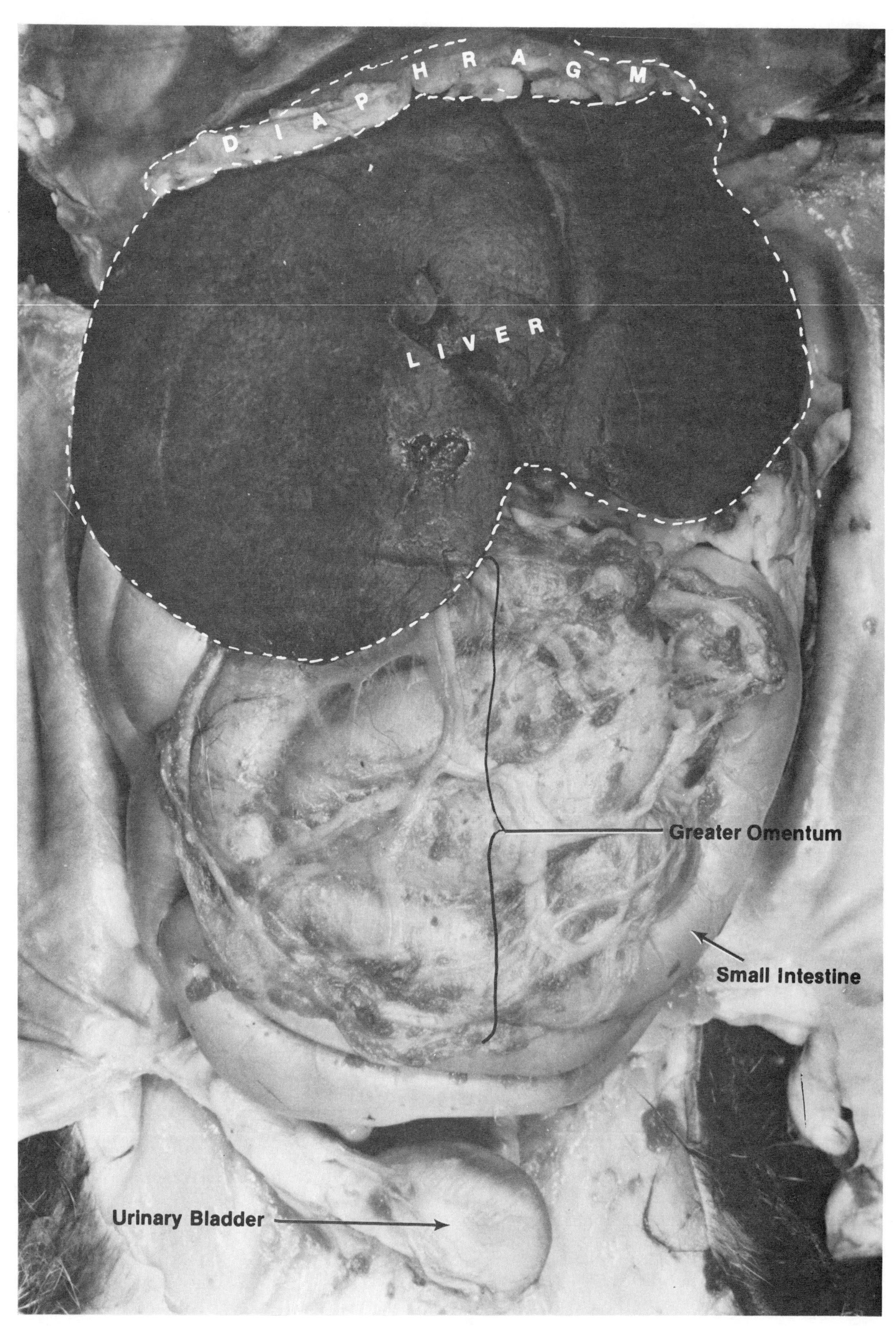


THE ABDOMINAL CAVITY (INTACT)

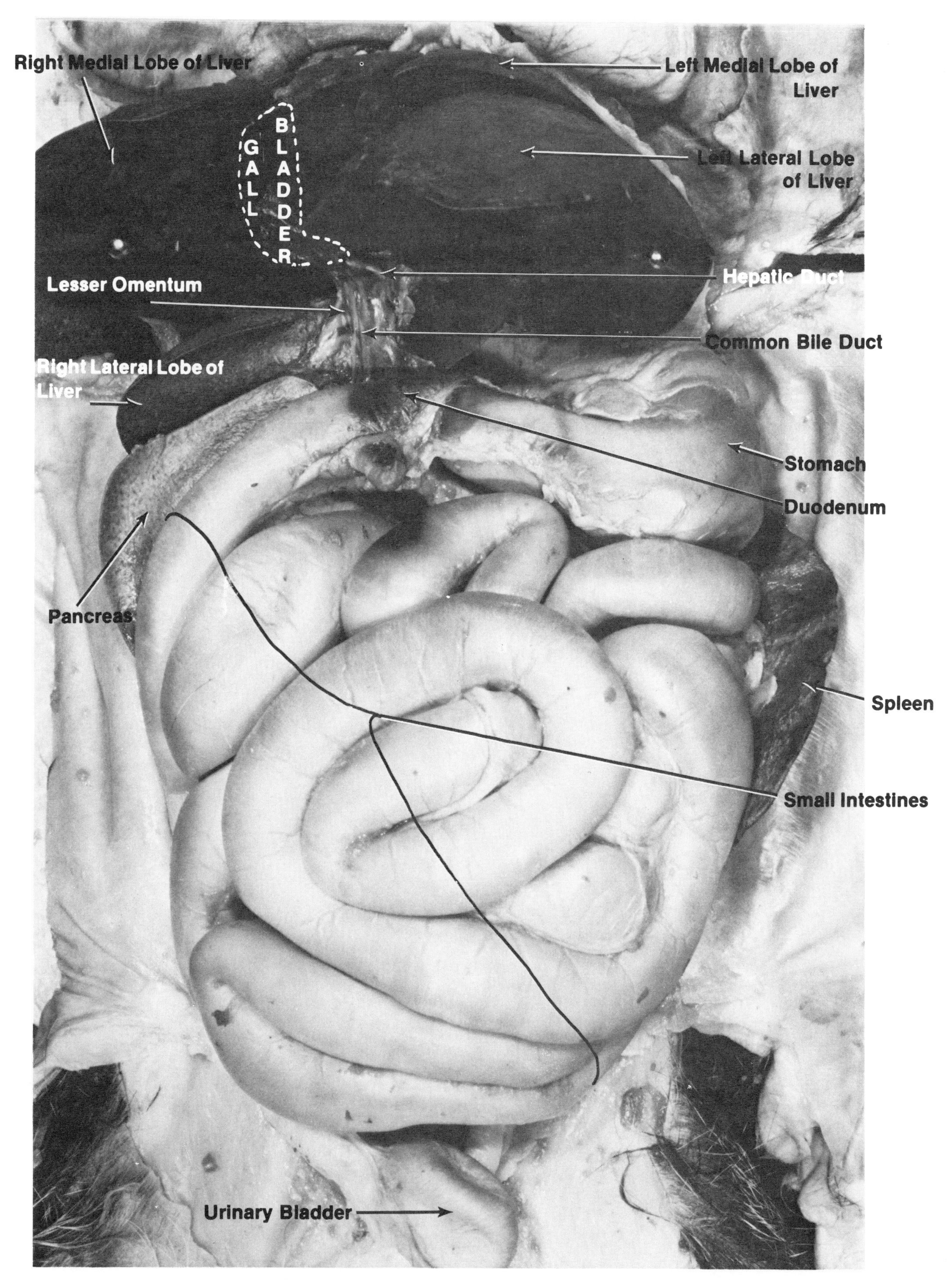


THE ABDOMINAL CAVITY (GREATER OMENTUM REMOVED)

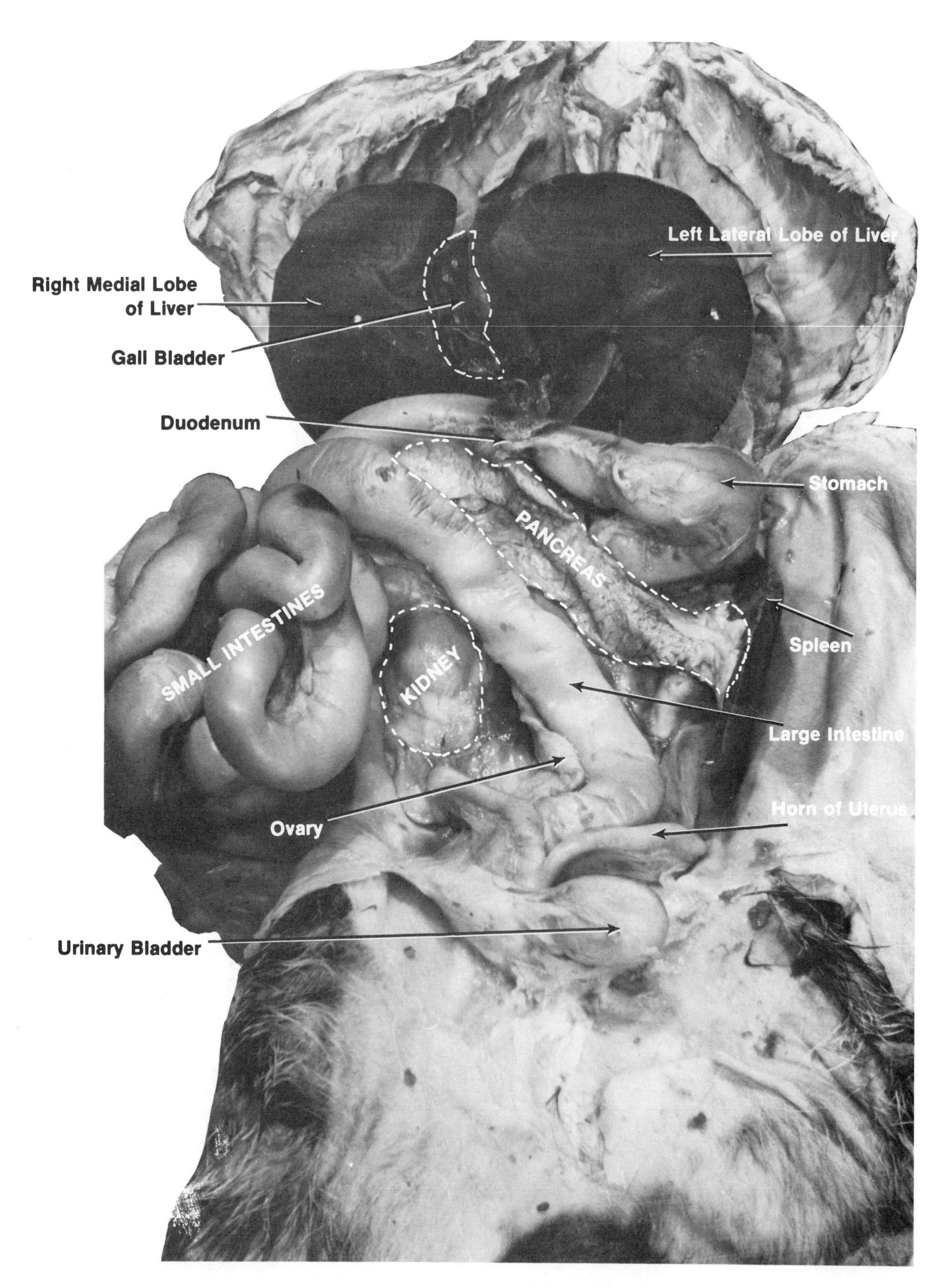


THE ABDOMINAL CAVITY (INTESTINES ASIDE)

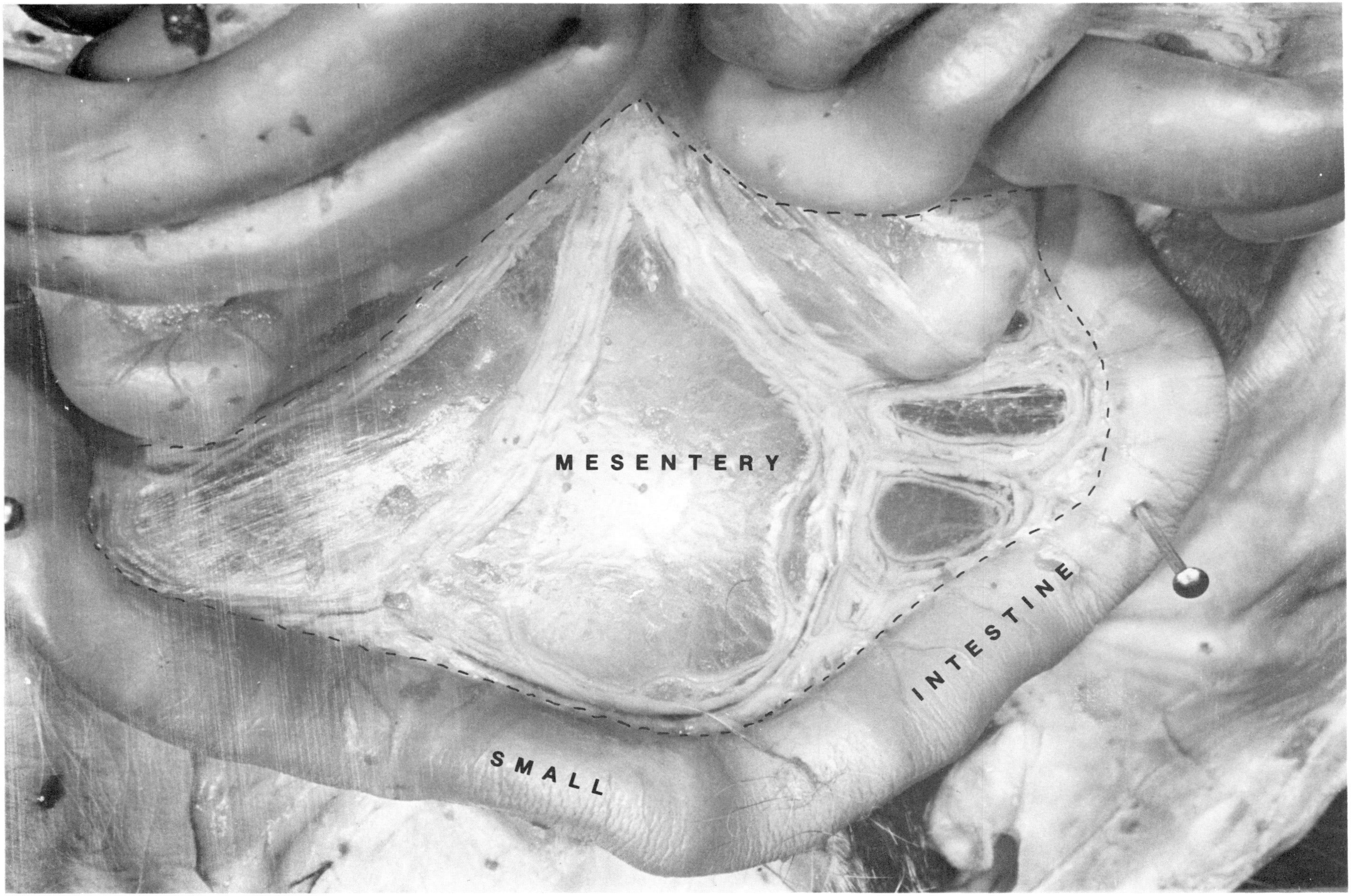

MESENTERY

SMALL INTESTINE

THE ABDOMINAL CAVITY - THE MESENTERY

Name ................................ Section ....... Date ................

# SELF - QUIZ III
# THE ABDOMINAL CAVITY

1. Name the five lobes of the liver.
2. Describe the structures of the greater and lesser omentum.
3. How do the jejunum and ileum differ?
4. What is the location and function of the ileocecal valve?
5. Name the parts of the large intestine beginning at its origin.
6. How does the location of the adrenal gland differ in the cat and in man?
7. How is the urinary bladder suspended?
8. Name the three major tubular structures that pass through the diaphragm?
9. Which of the following is not found in the abdominal cavity?
   a) gall bladder, b) kidney, c)trachea, d)pancreas

ANSWERS

1. ______________________________
2. ______________________________
3. ______________________________
4. ______________________________
5. ______________________________
6. ______________________________
7. ______________________________
8. ______________________________
9. ______________________________

10. Define, identify, and locate each of the structures listed below.
    a. esophagus ______________________
    b. duodenum ______________________
    c. jejunum ______________________
    d. colon ______________________
    e. fundus ______________________
    f. rugae ______________________
    g. pyloric sphincter ______________________
    h. mesentery ______________________
    i. cystic duct ______________________
    j. common bile duct ______________________

Label all of the features indicated on the photograph.

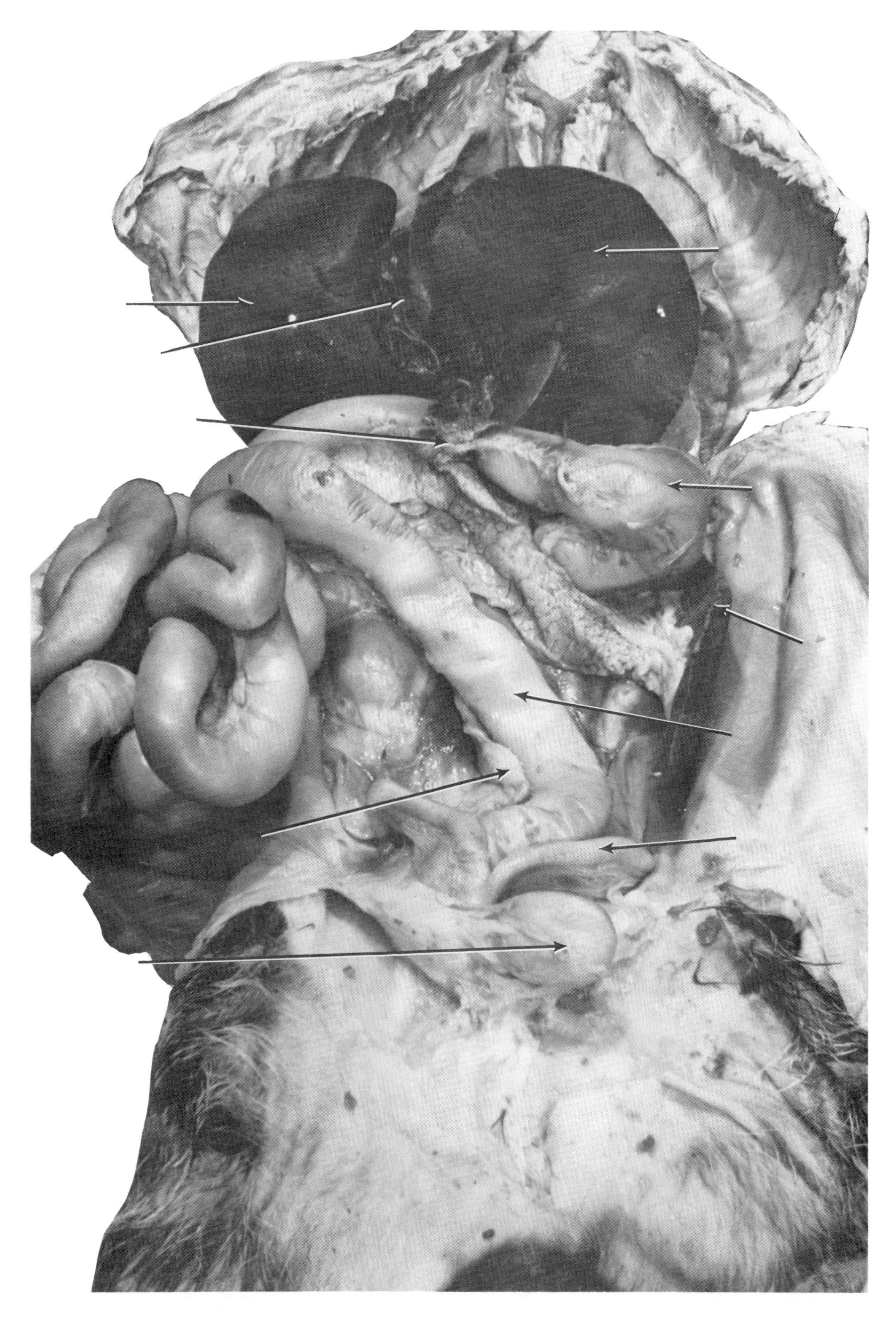

# THE THORACIC CAVITY

Begin your dissection of the thoracic cavity by making an incision with your scissors at the base of the rib cage about a half inch to the right or left of the mid-ventral line. This will avoid hitting the bony sternum and you will be cutting across softer costal cartilage. Continue your incision until the top rib has been cut. Spread the rib cage. Cut each rib near its dorsal origin to open the rib cage maximally.

**Diaphragm** - This muscular sheet forms the floor of the thoracic cavity. We have already studied it in the last section when discussing the abdominal cavity. Cut the diaphragm away from the body wall in a ventral to dorsal direction. The view at this point should be very similar to that in the photo.

**Pleura** - This is the serous membrane found within the thorax. The *parietal pleura* lines the inner walls while the *visceral pleura* covers the organs of the thorax.

**Lungs** - In the cat the right lung is composed of four lobes, the left of only three. In man the right lung has three lobes, the left has only two. They feel spongy when alive and more rubbery in preserved specimens. Each lung lies within a separate *pleural cavity*, the space between the lung and the thoracic body wall.

Cut a small, flat section of lung and observe with a hand lens or dissection microscope. If your specimen has been doubly injected (arteries and veins) you should observe three types of vessels within the lung tissue:

1. **Pulmonary Artery** - Branches of this vessel contain *blue* dye.
2. **Pulmonary Vein** - Branches of this blood vessel contain *red* dye.
3. **Bronchioles** - These branches of the bronchi, distributed throughout the lungs, are hollow with *white* edged walls.

**Mediastinum** - This is the area between the two pleural cavities. It extends from the sternum to the vertebral column and contains all of the thoracic viscera except the lungs

**Pericardium** - This double membrane encloses the heart. The *visceral pericardium* adheres closelt to the outer wall of the heart, while the *parietal pericardium* forms a sac which encloses the heart.

The pericardium is very prominent in the photo. As you can see, it encloses the heart as well as the large blood vessels entering and leaving the heart. Fat tissue is embedded within it. The *phrenic nerves* which innervate the diaphragm pass along the lateral edges of the pericardium. Identify these nerves.

**Thymus** - This gland is large in younger animals and becomes smaller as the animal matures. It lies ventrally upon the lower trachea and extends to the heart where it is associated with the pericardium.

The thymus and pericardium should be removed before studying the heart and thoracic blood vessels.

**Trachea** - This tube, the windpipe, extends along the mid-ventral portion of the neck extending into the thoracic cavity. Here it branches to form the two *bronchi* which penetrate the lungs. The air passage is always kept open by cartilage rings which stiffen and give a circular shape to the walls of the trachea. They are incomplete dorsally, thus forming the letter "C". The trachea lies ventral to the esophagus.

With your scissors cut out a half inch section of the trachea. Observe the rings. Slit the section longitudinally along the ventral side. Observe with a hand lens.

**Larynx** - This structure also known as the voice box was located at the top of the trachea earlier during the dissection of the oral cavity. Its uppermost segment is the *epiglottis*. It is associated with the *hyoid bone* anteriorly. It is composed of several cartilaginous elements. These are the:

> **Epiglottis** - a cone-shaped flap of cartilage located at the top of the larynx. (See page 74 for a more detailed description.)
> **Thyroid Cartilage** - anterior, large, shield-shaped
> **Cricoid Cartilage** - posterior, ring-shaped
> **Arytenoid Cartilages** - paired, small, pyramid-shaped, located dorsally at cranial border of cricoid cartilage

With your scalpel make a longitudinal incision into the mid-ventral surface of the larynx. Continue to cut until the larynx has been divided into right and left halves. You have now exposed the vocal cords. These are folds of mucous membrane, supported by elastic ligaments which extend from the arytenoid cartilages to the thyroid cartilage near the base of the epiglottis. Two such membrane pairs are noted, the more anterior *false vocal cords* and the posterior *true vocal cords*. It is the latter pair that produce sound. The *glottis* is the space between the true vocal cords.

**Thyroid Gland** - This gland is composed of two elongated *lobes* which lie on either side of the trachea right below the larynx. They extend for a distance of about ¾ inch. A connecting segment between the two lobes, the *isthmus*, is very narrow and may be destroyed during the dissection. Two pairs of *parathyroid glands* are embedded in the dorsal surface of the thyroid. They cannot, however, be seen without microscopic examination.

## HUMAN RESPIRATORY TRACT

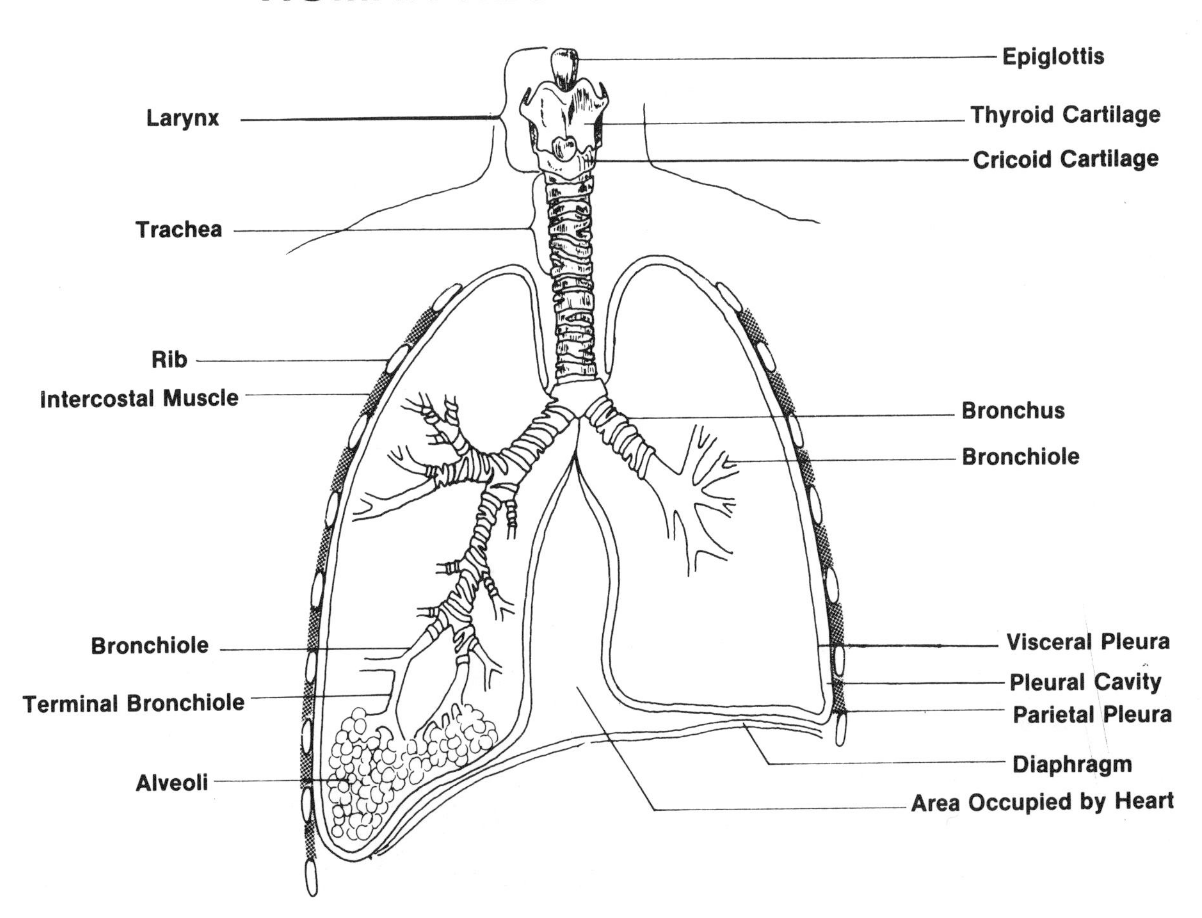

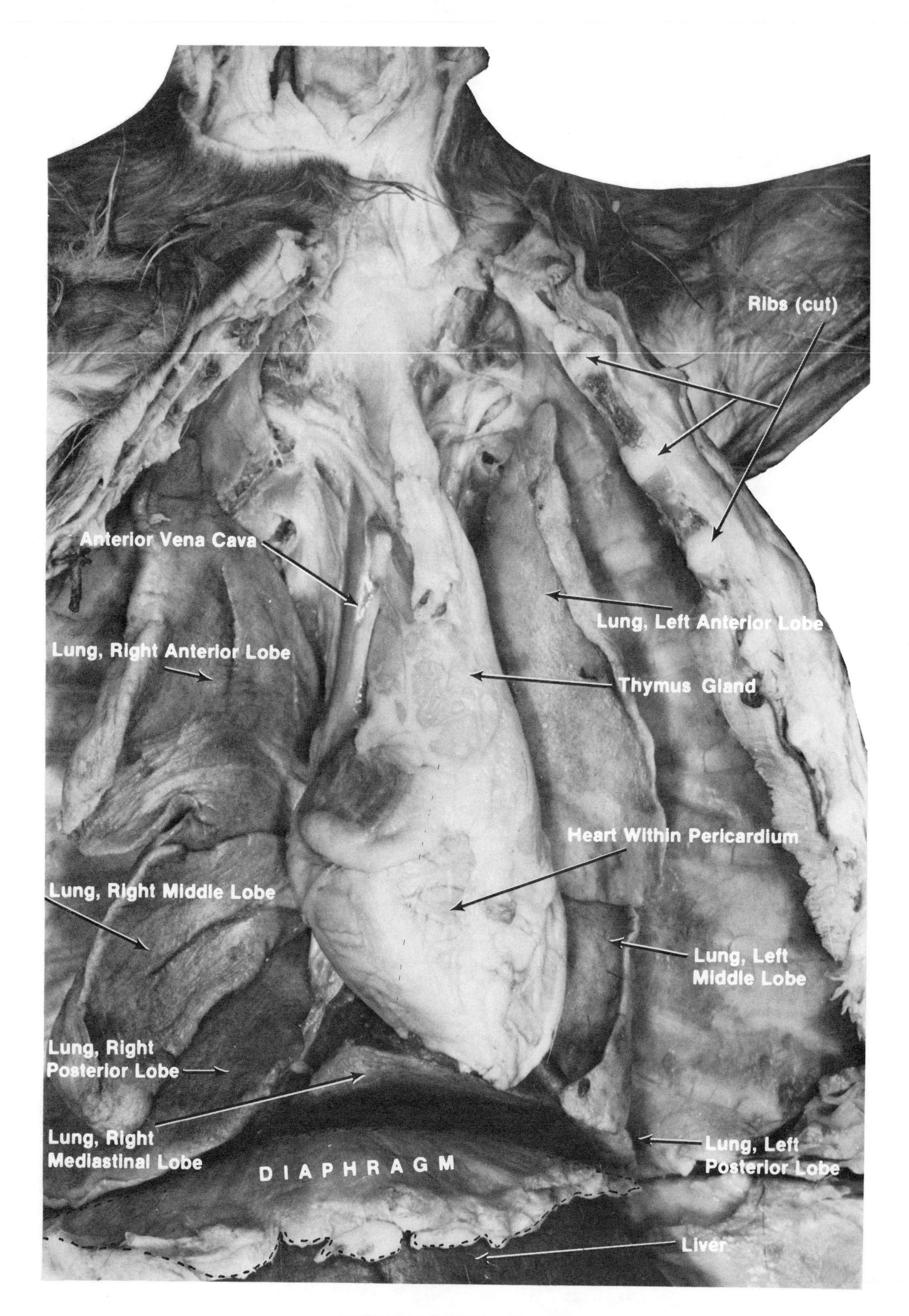


THE THORACIC CAVITY

Name .................................. Section ........ Date ................

# SELF - QUIZ IV
# THE THORACIC CAVITY

1. Name the seven lobes of the cat lung.
2. How many pairs of ribs does the cat possess? man?
3. What three types of vessels are seen when we section the lung?
4. Describe the structure of the pericardium.
5. Do the bronchi contain cartilage rings? the bronchioles? the alveoli?
6. We inhale 20% oxygen and 0.04% carbon dioxide. What percentages of the exhaled air are oxygen and carbon dioxide?
7. Name the cartilaginous structures that make up the larynx.
8. Name the hormones secreted by the thyroid and their functions.
9. What is the function of the thymus gland?
10. Define each of the terms listed below.

ANSWERS

1. ______________________________
2. ______________________________
3. ______________________________
4. ______________________________
5. ______________________________
6. ______________________________
7. ______________________________
8. ______________________________
9. ______________________________
10. a. tidal volume ______________________________
    b. emphysema ______________________________
    c. pneumonia ______________________________
    d. surfactant ______________________________
    e. mediastinum ______________________________
    f. phrenic nerve ______________________________
    g. Hering-Breuer reflex ______________________________
    h. negative pressure breathing ______________________________
    i. pleurisy ______________________________
    j. parathyroid glands ______________________________

Label all of the features indicated on the photograph.

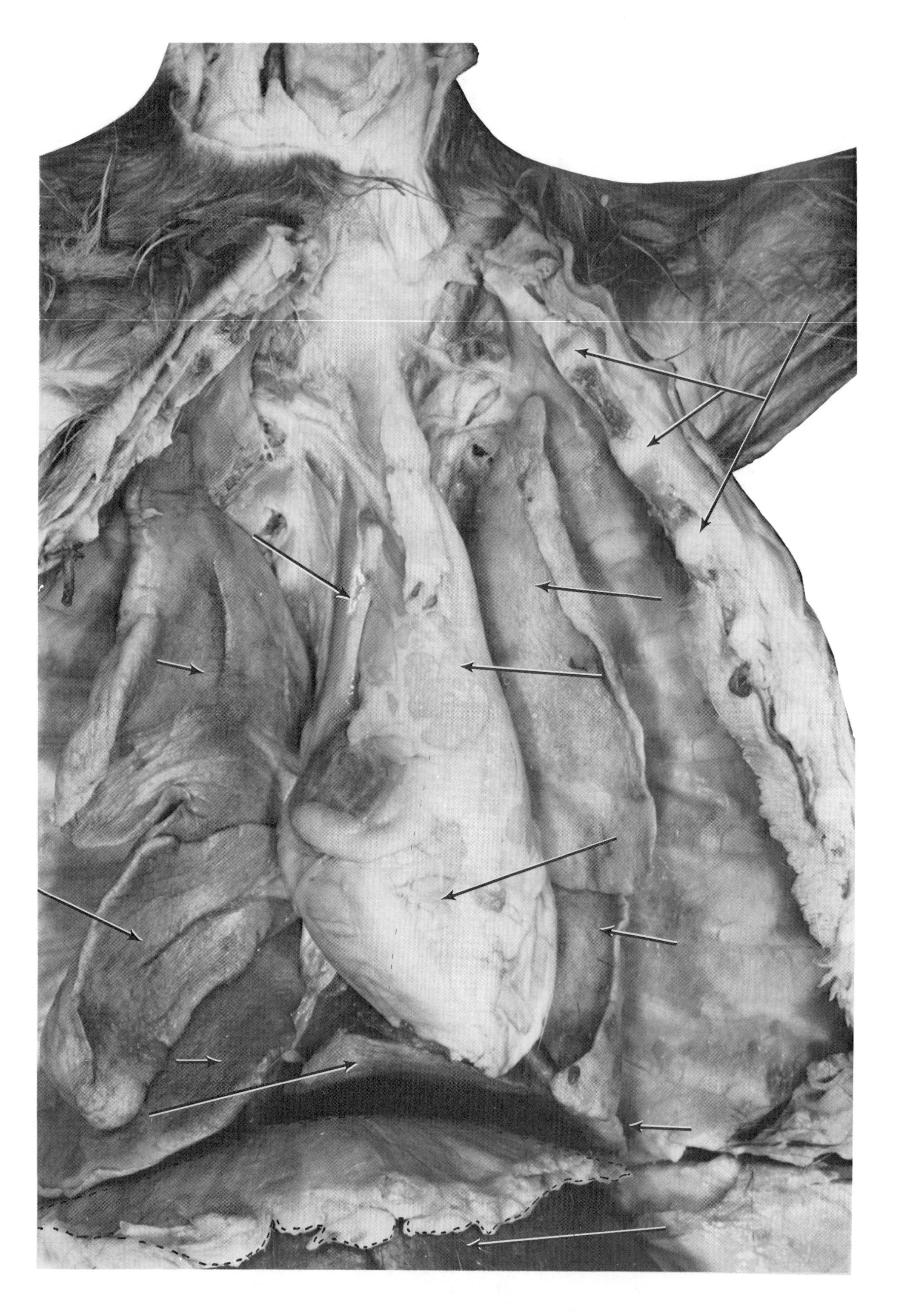

# SHEEP HEART

## VENTRAL VIEW

The sheep heart is studied because it is larger than the cat's heart and closer in size to that of the human.

Examine the sheep heart and note the conical shape. The tip of the heart, or the apex, is the most posterior section. Much of the fat surrounding the blood vessels at the anterior end of the heart has been removed to facilitate observation. Almost all of the *parietal pericardium*, the fibrous serous membrane enclosing the heart has been removed. In some specimens parts of it will still be found fused to the bases of the large blood vessels. The *epicardium*, or outer layer of the heart, is covered by the finer visceral pericardium

Hold the heart upright as in the photo, with most of the large vessels opening to the rear. You are now looking at the ventral side of the heart. The right side of the heart is at your left hand; the left side of the heart is at your right hand.

Find and identify the ear-like structures atop the ventricles. These are the *right* and *left atria*. (The term auricle is sometimes used. It means "little ear".)

The *pulmonary artery* is the prominent blood vesel leaving the heart at the right ventricle in the upper mid-ventral area. It passes anteriorly toward the left.

The largest and widest artery of the body, the *aorta*, can be seen next to its first major branch, the *brachiocephalic* artery. This vessel carries blood to the right shoulder and arm as well as to the head.

One of the *coronary arteries* may be seen along the ventral surface of the ventricle. It lies within the ***longitudinal sulcus***, a depression upon the surface corresponding to the line of separation between the right and left ventricles. The *coronary veins* generally run along with the arteries.

## DORSAL VIEW

Now turn the heart around and observe the dorsal surface. This time the right and left sides of the heart correspond to your own.

Find the *right atrium* and *left atrium*. The *pulmonary* artery which we saw in the last photo is seen here divided into two smaller tubes, the *right* and *left branches*. One passes to each lung.

The wide and thick-walled *aorta* may also be seen along with the *brachiocephalic artery*.

Not visible in the previous photo but clearly seen here are:

- The *superior vena cava* which empties into the right atrium carrying deoxygenated systemic blood from the head, arms, and upper chest.
- The opening of the *inferior vena cava* which carries deoxygenated systemic blood from the lower parts of the body
- The entrance of one of the *pulmonary veins* into the left atrium carrying oxygenated blood from the lungs.

Find each of these structures in your specimen.

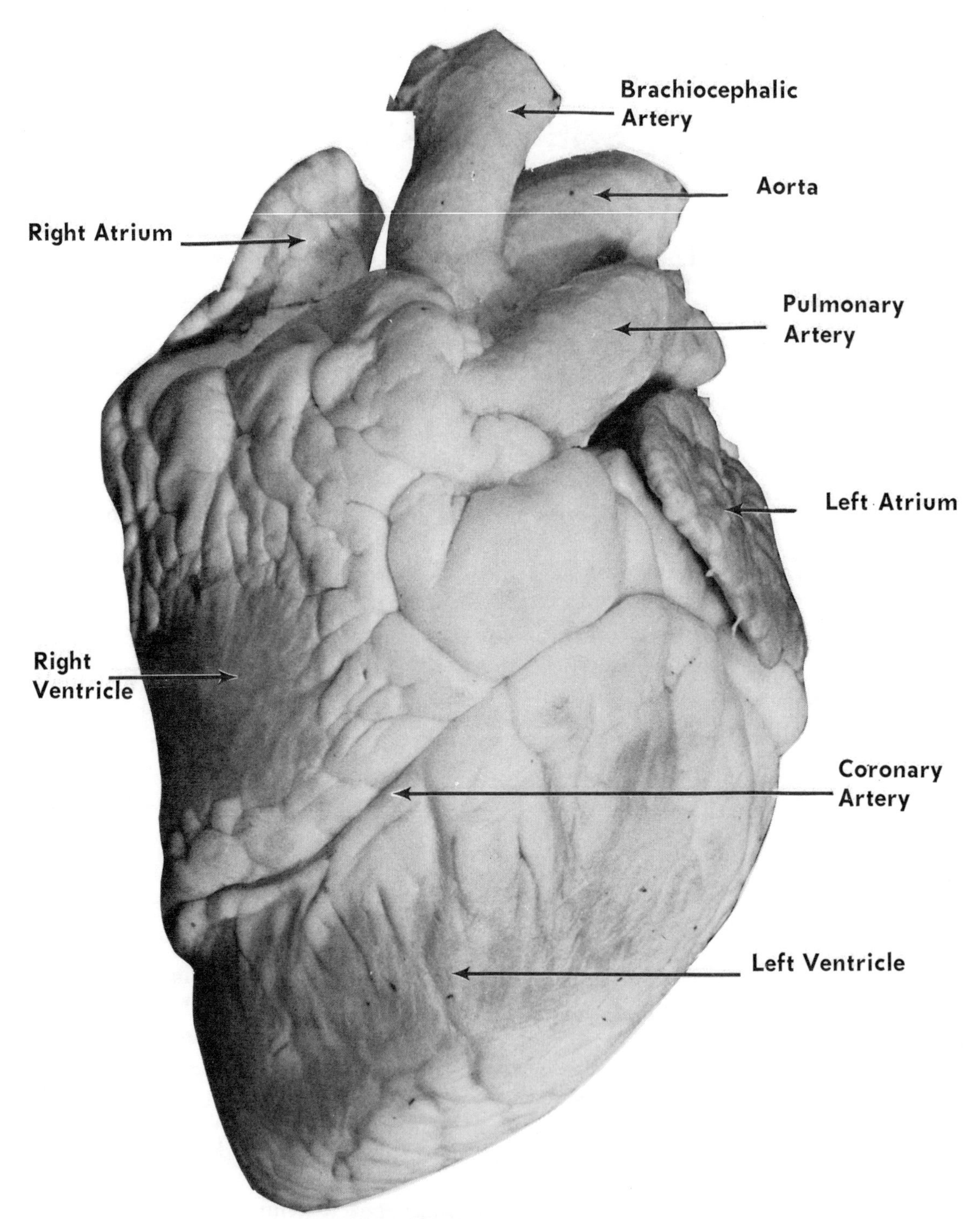


SHEEP HEART, VENTRAL VIEW

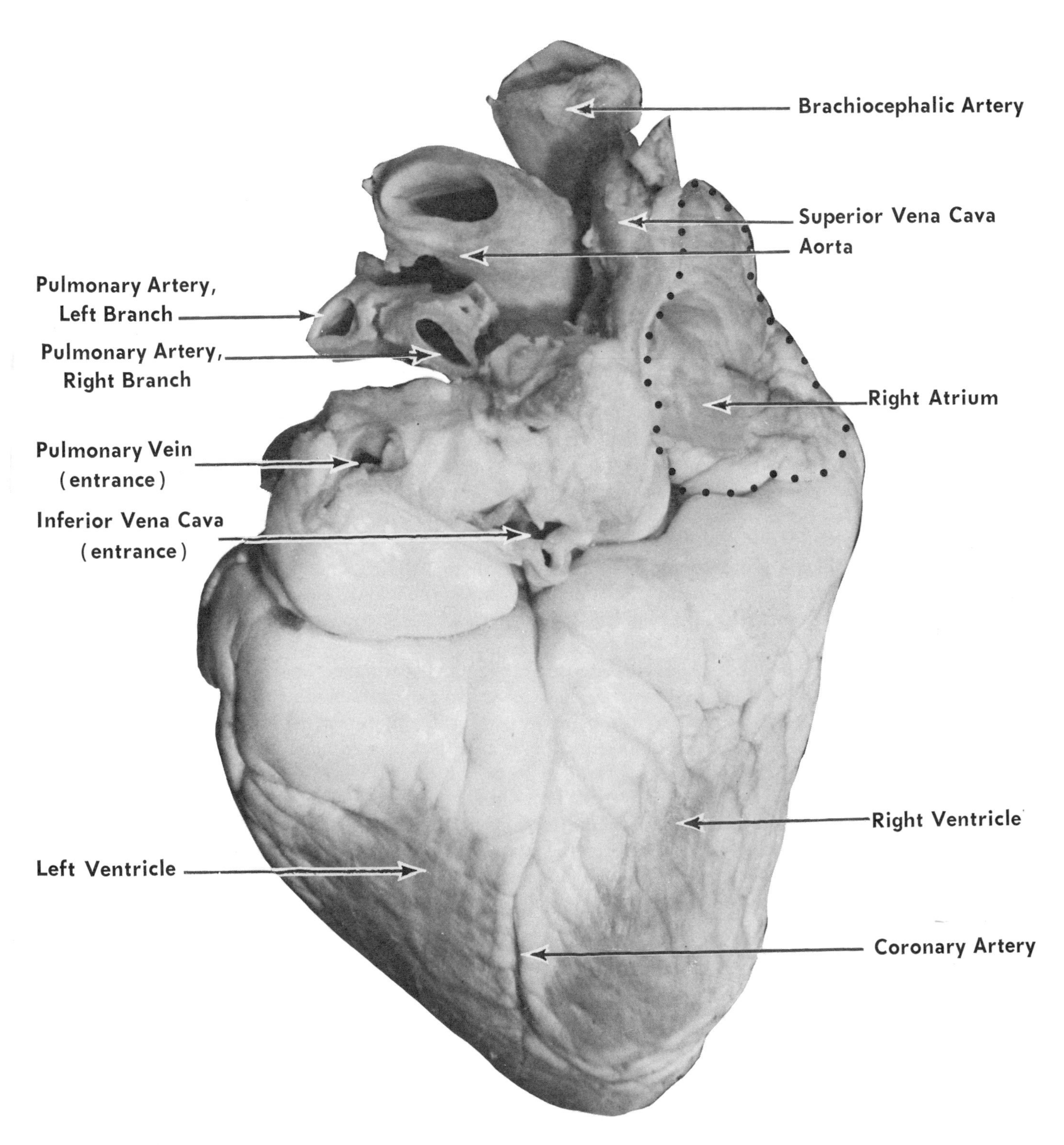


SHEEP HEART, DORSAL VIEW

# SHEEP HEART (OPEN), CORONAL PLANE

Note: At the start of the sheep heart dissection divide the class in two. Half will follow the dissection procedures outlined here, the other half will dissect as in the following two photos. The groups should then exchange specimens with one another.

Place the sheep heart in a dissection pan, ventral surface up. Use your scalpel to cut into the *myocardium*, the muscle layer that comprises the major portion of the heart. Begin your incision at the posterior edge of one of the atria and continue downward to the *apex*, or tip of the heart. Then continue upward on the second side to the posterior border of the other atrium. Separate the two ventricles of the heart as in the photo, leaving the atria intact.

The upper part of the picture shows the inside of the ventral half of the sheep heart, the lower part shows the dorsal half.

Observe the thickness of the outer muscular wall (myocardium). Can you tell which is the right and which is the left side of the heart? The thickness of the wall will reveal this. To what parts of the body does the left ventricle pump blood? To what parts of the body does the right ventricle pump blood?

Note the *chordae tendineae.* At one end these tough, white tendinous "heart strings" are attached to the *atrioventricular* values (A-V valves). To what are they attached at the other end? Their function is to prevent the A-V valves from being turned inside out during the high pressure phase of ventricular systole. A backflow of blood into the atria is thus prevented.

Note the very musclular *interventricular septum,* simply labeled "septum" in the photo, a dividing wall between the right and left ventricles. It effectively separates oxygenated and deoxygenated blood in the ventricles and participates in the pumping action of the heart. A much thinner *interatrial septum* is found between the two atria.

Also note the large blood vessels entering and leaving the heart. Use wooden probes to trace these vessels. In which chamber of the heart does each originate or terminate? Can you identify these vessels?

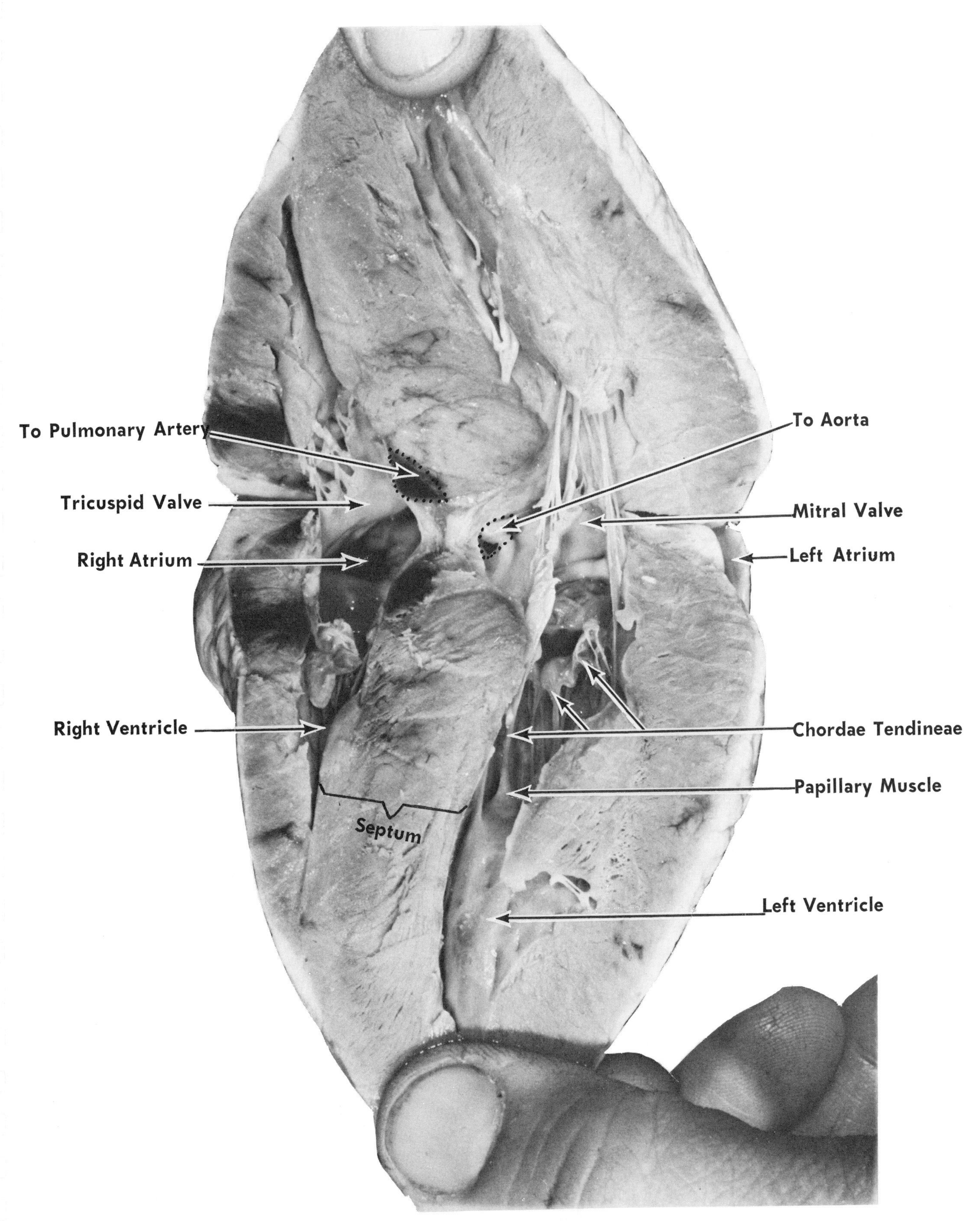


SHEEP HEART (Open), CORONAL PLANE

# SHEEP HEART (OPEN)

## RIGHT SIDE

In order to expose the right side of the heart as in the photo, proceed as follows:

Open the *pulmonary artery* by cutting longitudinally along its length. Continue to cut through the *myocardium* of the *right ventricle* to the apex of the heart. Similarly, cut open the *superior vena cava.* Continue the cut through the wall of the *right atrium* until the two cuts cross. Now, spread apart the heart and observe.

Note:

—The semilunar valve at the origin of the *pulmonary artery.*
—The tree cusps, or flaps of tissue, of the *tricuspid valve* and the tough *chordae tendineae* holding the valve, in place.
—The *papillary muscles* from which the chordae tendineae originate.
—The *superior vena cava* entering the *right atrium.*
—The inner rough textured walls of the right atrium.
—The entrance for the *inferior vena cava.*
—The opening of the *coronary sinus* bringing blood from the coronary veins and the flap-like valve at the entrance.

Compare the thickness of the walls of the atrium and ventricle. Relate these to their functions.

Find all of the structures labeled in the photo.

## LEFT SIDE

You may use the same specimen of sheep heart as in the preparation seen in the last photo.

Turn the heart to the left side. Find the *aorta* and open it by cutting longitudinally along its length. Continue your incision posteriorly through the *myocardium* of the *left ventricle* to the apex of the heart. Then, cut open the left atrium along its lateral surface. Now, spread the heart apart and observe.

Note:

—The *semilunar valves* at the base of the aorta.
—The openings in the wall of the aorta where the *coronary arteries* and the *brachiocephalic artery* originate.
—The two cusps, or flaps of tissue, of the *bicuspid (mitral) valve,* the *chordae tendineae* and the *papillary muscles.*
—The inner surface of the *left atrium* and the opening of the *pulmonary veins.*
—The *pulmonary artery* may also be seen.

Find all of the structures labeled in the photo.

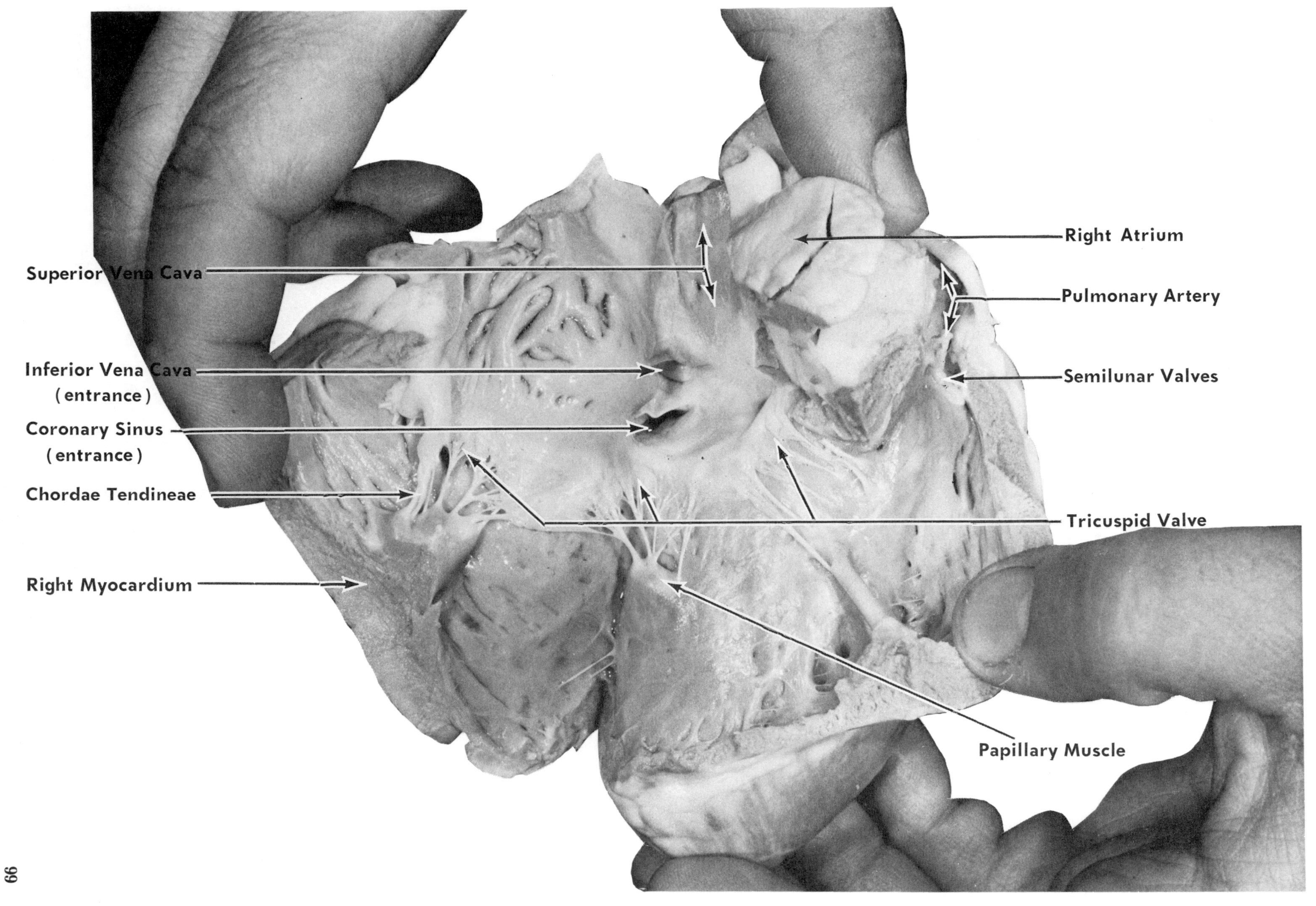


SHEEP HEART (Open), RIGHT SIDE

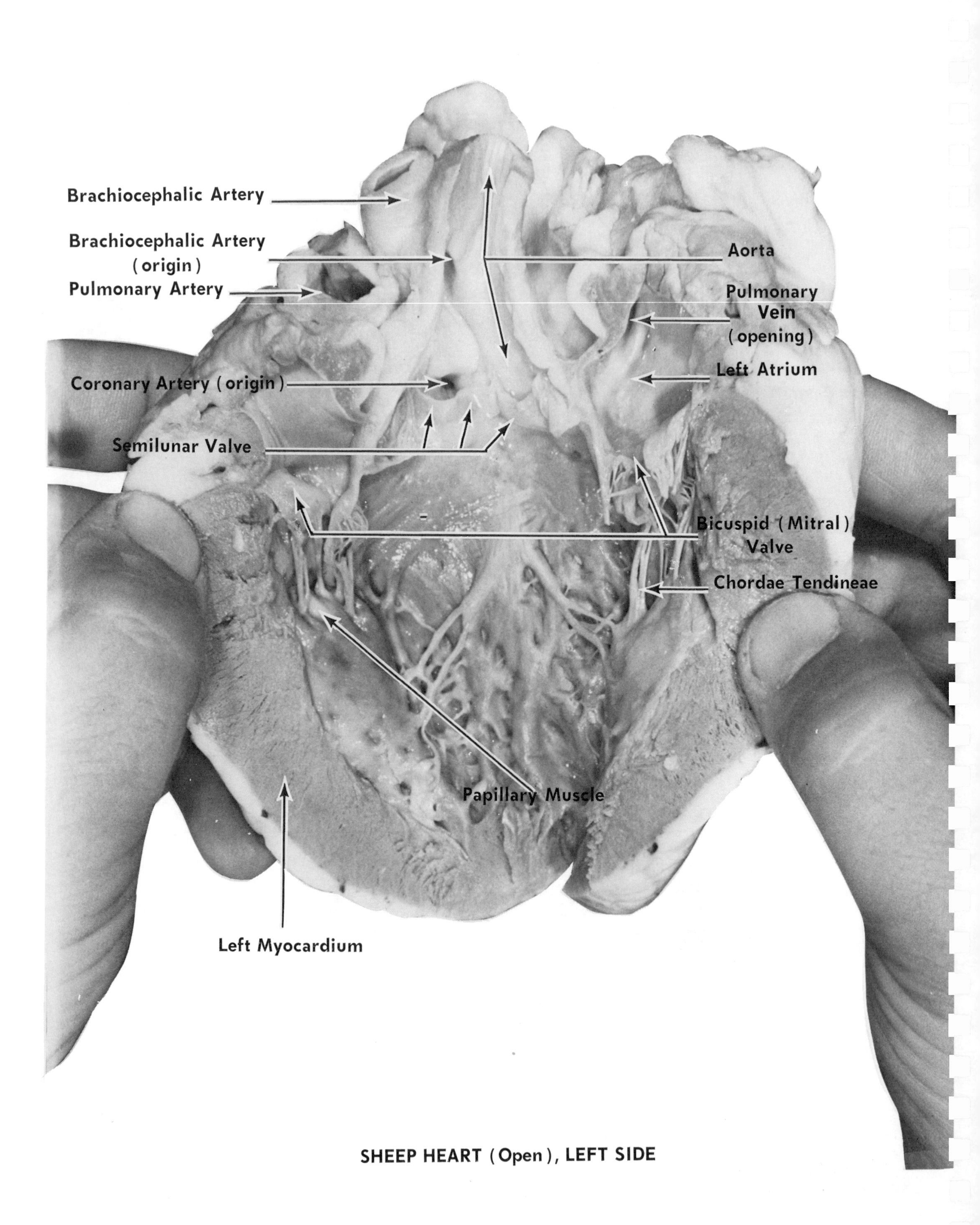


SHEEP HEART (Open), LEFT SIDE

# THE HUMAN HEART

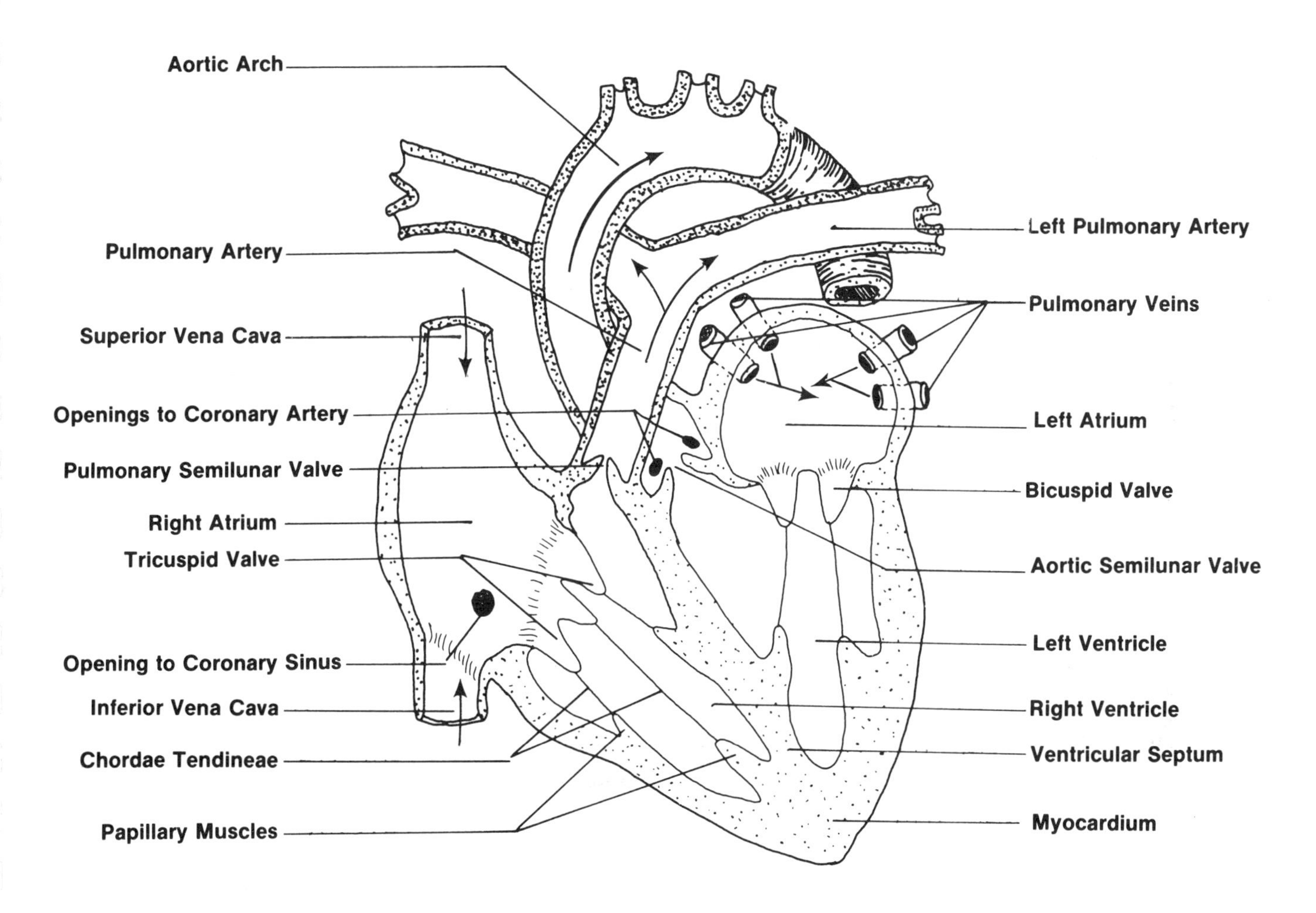


Note: Arrows Indicate The Path Of Blood Flow

Name ............................ Section ............ Date ..................

# SELF-QUIZ VI
# THE HEART

1. Name the membrane sac which encloses the heart.
2. A large gland which extends into the neck region partially covers the heart. Name this gland.
3. Another name for the mitral valve is the .............. valve.
4. Name the valves located at the origin of the aorta.
5. The first branches of the aorta, located near its base, lead to the ........... and are known as the ............... arteries.
6. Blood returning from the lower portion of the body enters the right atrium of the heart through the blood vessel known as the ........................................
7. The flaps of the atrioventricular valves are held in place by the tough fibrous ............
8. Oxygenated blood from the lungs enters the heart at the (a) right atrium, (b) left atrium, (c) right ventricle, (d) left ventricle.
9. The right and left ventricles are separated by the ..................................
10. Define each of the terms listed below.

ANSWERS

1. ____________________
2. ____________________
3. ____________________
4. ____________________
5. ____________________
6. ____________________
7. ____________________
8. ____________________
9. ____________________
10. a. papillary muscle ____________________
    b. foramen ovale ____________________
    c. endocardium ____________________
    d. coronary thrombosis ____________________
    e. fibrillation ____________________
    f. mitral stenosis ____________________
    g. heart murmur ____________________
    h. rheumatic heart ____________________
    i. chordae tendineae ____________________
    j. tachycardia ____________________

Label all of the features indicated on the photograph.

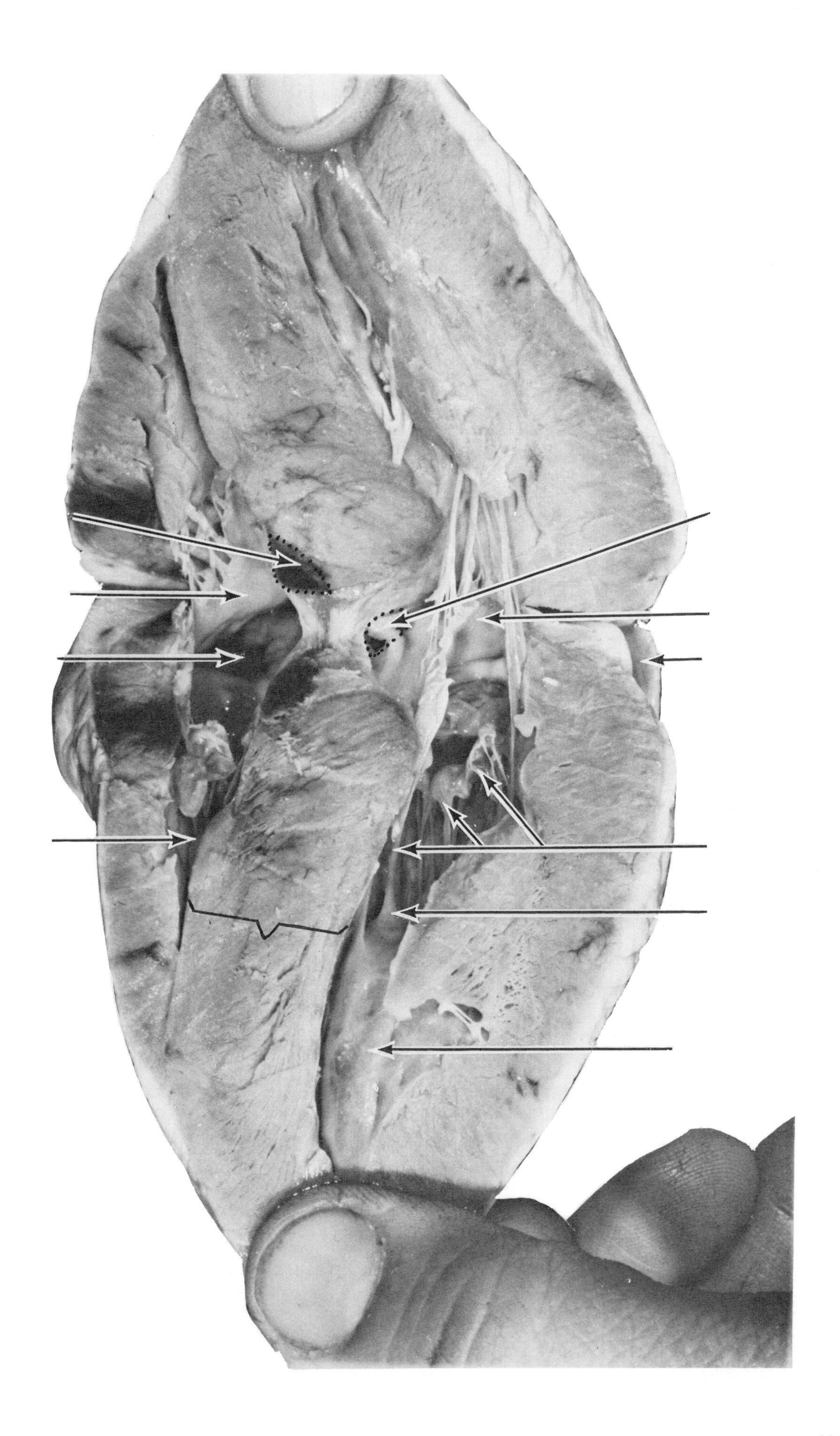

# VEINS - ANTERIOR REGION

We begin our study of the major blood vessels with the veins which carry blood to the heart from the head, forelimbs, shoulder, and thoracic region.

The anterior veins lie close to the surface of the body and are therefore studied before the arteries which lie deep to the veins. In doubly injected specimens veins will be blue, arteries will be red. This is due to the latex dye with which they were injected. This permits you to see these vessels more clearly and to differentiate between arteries and veins.

You have already exposed the heart and some of the major veins in your study of the thoracic cavity. Extend your incision into the neck area. If you have not yet done so, expose the *trachea, larynx*, and *thyroid gland* as in the photo. Remove the *thymus gland* carefully from the ventral surface of the heart and lower trachea.

Your best dissection instrument at this point is the dissecting needle. Use it instead of a scalpel or scissors to expose and follow the path of a blood vessel. Scalpels and scissors, especially in the hands of the novice tend to destroy rather than preserve or expose. It is used for clearing a blood vessel of the connective tissue adhering to it, to tear into a thin muscle layer to follow a vessel, to separate one blood vessel from another, or from the nerves associated with it.

Proceed to expose the large veins surrounding the heart.

**Anterior Vena Cava** - Locate the trunk of this major systemic vein above the heart. In the cat the term anterior vena cava is more correct and therefore preferred to superior vena cava. All anterior veins lead into the anterior vena cava. Are there any exceptions? Where does the anterior vena cava empty?

**Posterior Vena Cava** - This is the major vein returning blood from the lower extremities and from the abdominal area. Again, this designation in the cat is preferred to inferior vena cava. It can clearly be seen rising from the diaphragm, which it has penetrated, to enter the heart at the right atrium together with the anterior vena cava.

**Innominate** (Brachiocephalic) - Follow the anterior vena cava cranially from the heart. It divides into two equal branches, the "V" shaped *innominate veins* also known as the brachiocephalic veins. Each of these is formed by the union of the *subclavian vein* from the shoulder and the *external jugular vein* from the neck.

**Jugular Veins** - Two jugular veins lie along each side of the neck. The larger *external jugular* drains the head and neck, while the *internal jugular vein* drains the brain and spinal cord. It empties into the external jugular vein.

**Subclavian** - This vein is a continuation of the *axillary vein* in the shoulder joint area. The axillary vein is a continuation of the *brachial vein* in the arm.

**Internal Mammary** (Sternal) **Vein** - On the ventral surface of the anterior vena cava about one inch above the heart the *internal mammary* or sternal vein arises. It results from the union of the paired internal mammary veins which drain the ventral thoracic body wall and the mammary glands of female cats.

**Azygous** - Push the right lung and the heart medially. On the dorsal body wall, to the right of the midline, and running parallel to it, observe a vein which enters the anterior vena cava near the heart. Only a very small portion of it is seen in the photo. This unpaired vessel is the *azygous vein.* Carefully dissect its transverse tributaries, the *intercostal veins,* one pair between each two ribs. It drains the dorsal body wall and the dorsal musculature.

**Jugular Veins** (continued) -Continue to trace the *external jugular vein* cranially. At the level of the masseter, the cheek muscle, it divides to form the *posterior facial vein* and the *anterior facial vein.* There is a third branching of the exteranl jugular slightly anterior to this, at the upper end of the larynx, forming the *transverse jugular vein.* It crosses the neck ventrally to connect the right and left external jugular veins. Near the ear the posterior facial vein divides to form the *anterior auricular vein* and the *posterior auricular vein.*

**Cephalic** - This vein was seen on the lateral surface of the foreleg when the cat was skinned. It is a long vein which continues up to the shoulder area and there joins the *transverse scapular vein.* This vein, in turn, enters the external jugular vein (see photo). The *cephalic* and *brachial vein* are joined in the inner elbow by the *median cubital vein.*

Other veins in the photo include the:

**Subscapular** - This vein arises at the underside of the scapula and enters the subclavian vein close to the base of the external jugular vein.

**Submental** - This is a branch of the anterior facial vein it drains the mandible.

**Long Thoracic** - This vein arises along the inner surface of the pectoralis minor muscle then passes to the latissimus dorsi. It joins the *axillary vein.*

**Thoracodorsal** - This vein arises from under the scapula and similarly joins the *axillary vein.*

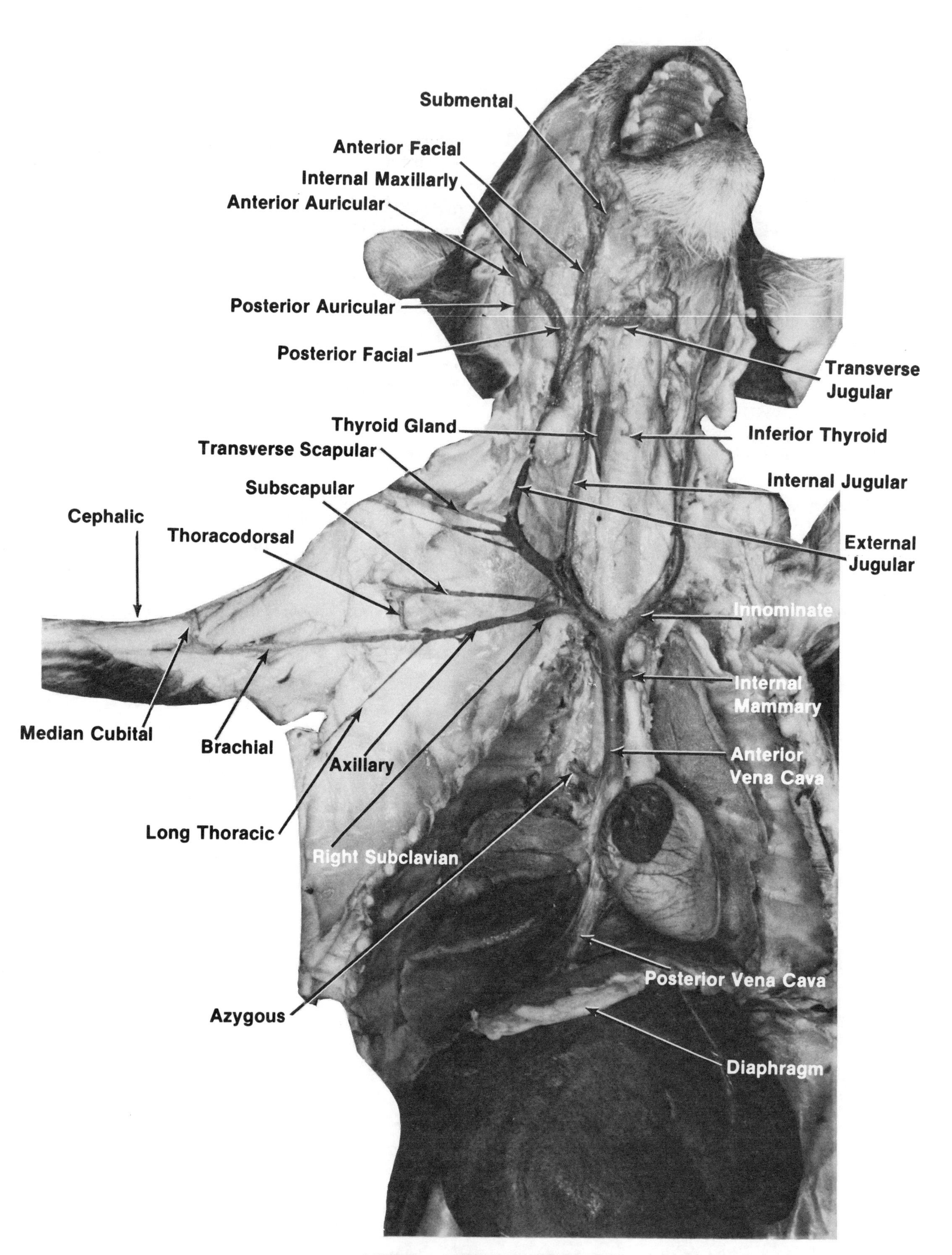


VEINS - ANTERIOR REGION

# VEINS - POSTERIOR REGION

The systemic veins below the heart empty into the *posterior vena cava.* The upper portion of this enlarged blood vessel was observed when we studied the thoracic veins. It passes through the diaphragm and lies along the mid-dorsal body wall of the thorax. It enters the right atrium of the heart near the entrance of the anterior vena cava.

**Posterior Vena Cava** - The major branches of this large vein will be described in descending order, in a caudal direction. Within the diaphragm it receives the *phrenic veins.* Below the diaphragm the posterior vena cava passes through the substance of the liver within which it receives the *hepatic veins.* Below the liver the following veins drain into the posterior vena cava:

**Adrenolumbar** - This vein drains the adrenal gland and the dorsal lumbar body wall.

**Renal** - Remove some of the fat covering the kidney and its vein, the *renal vein.* Most of the blood in this vein has been cleared of nitrogenous wastes.

**Right Spermatic or Ovarian** - This vein from the right testes or ovary enters the posterior vena cava directly. The *left spermatic or ovarian vein* enters the left renal vein instead. See these differences in your specimen. In females the vein is easier to find by first locating the ovaries and tracing the vein to where it enters the posterior vena cava.

Note: the last photo in this section, on page 110, shows these last three veins in great detail. It is a photo of a pregnant cat. That will explain the enlarged *ovarian vein.*

**Iliolumbar** - This vein drains the small of the back.

**Common Iliac** - In the pelvic area the posterior vena cava divides sharply into two veins that drain the legs, lower pelvis, and tail. They are the *common iliac veins.* They are formed by the *external iliac vein* and the much shorter *internal iliac vein.*

**Internal Iliac** - The tributaries of the internal iliac vein are the:

**Hypogastric** - This is sometimes called the *umbilical vein.* It comes from the lateral wall of the urinary bladder.

**Hemorrhoidal** - This vein arises from the region of the pubic symphasis and rectum.

**Gluteal** - These veins drain the gluteal area.

**External Ilaic** - The external iliac receives the *caudal* or *sacral vein,* from the tail region.

**Femoral** - In the thigh, the *external iliac* is known as the *femoral vein.* It receives several branches.

**Deep Femoral** - It drains the deeper muscles of the thigh.

**Muscularis** - It too drains muscles of the thigh.

**Superior Articular** - This vein arises in the region of the knee cap. It extends up the thigh to join the femoral vein.

**Greater Saphenous** - This long superficial vein lies on the muscle surface right below the skin. It comes up along the inner side of the calf of the leg and extends along the lower part of the gracilis muscle to join the femoral vein.

**Popliteal** - This vein comes from the deep, lower, posterior region of the thigh across the popliteal space (at the back of the knee). Remove the fat in the popliteal space to expose this vein.

**Anterior Tibial** and **Posterior Tibial** - These veins from the lower hind leg join the popliteal vein.

**Label the parts of the human venous system below and compare them with those you observed in the cat.**

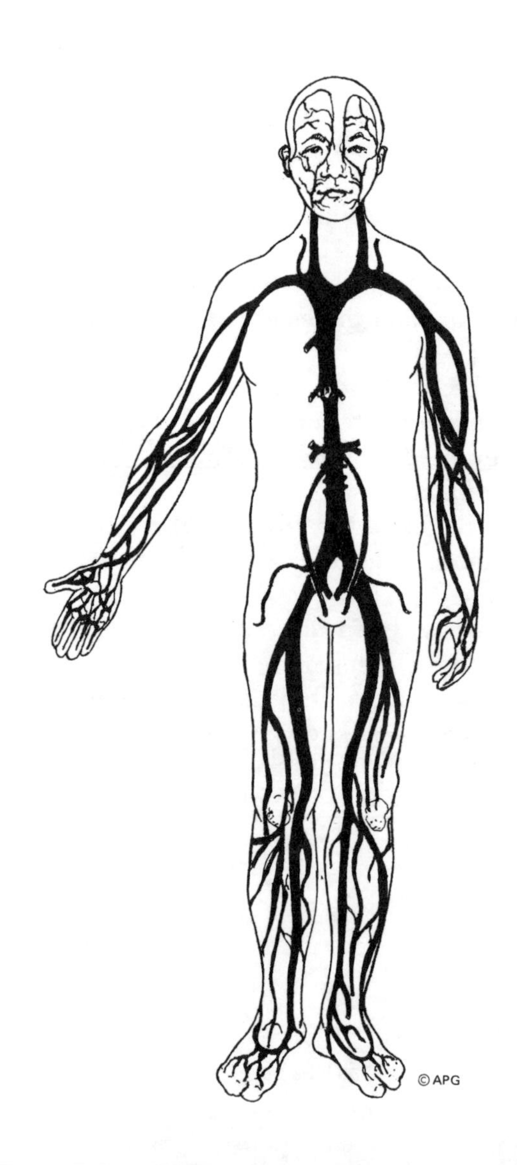

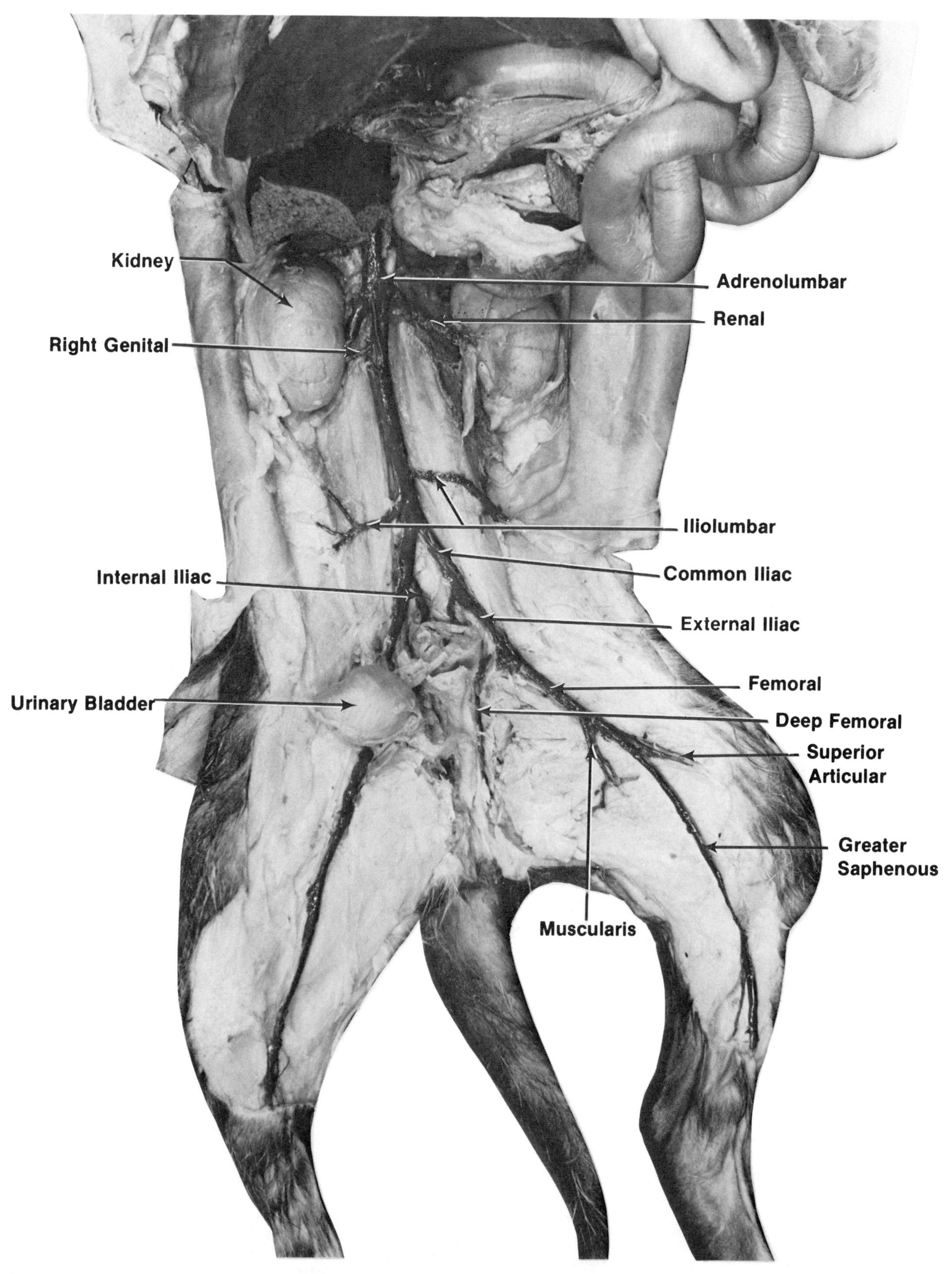


VEIN - POSTERIOR REGION

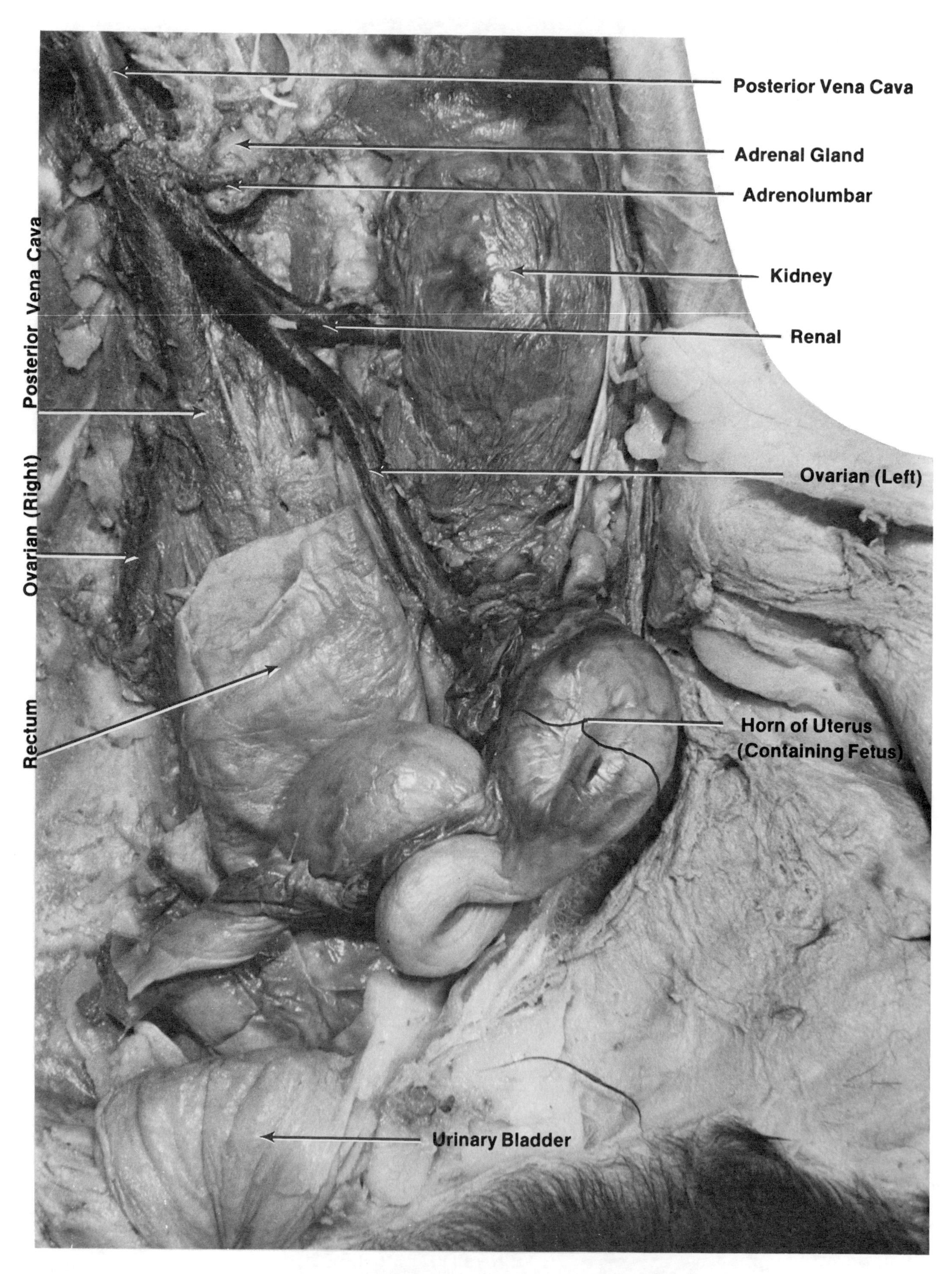


VEINS - POSTERIOR REGION (CLOSE-UP)

# VEINS - HEPATIC PORTAL SYSTEM

The veins from most of the abdominal viscera do not join the posterior vena cava directly. Instead, blood from the stomach, pancreas, spleen, small intestine, and large intestine drain into a larger vein known as the *hepatic portal vein* which enters the liver and there breaks up into capillaries called *sinusoids*. The hepatic portal vein is rich in digested nutrients. Before these enter the general circulation the liver transforms them according to the needs of the body. Glucose may be stored as glycogen, amino acids deaminated, or fatty acids may be converted to carbohydrates.

Note: In the accompanying photo, the stomach, spleen, pancreas, small, and large intestines have been pushed to the left in order to better expose the hepatic portal blood vessels.

The *hepatic veins* from the liver join the *posterior vena cava* within the substance of the liver. Thus the blood from the abdominal viscera is once again returned ot the general circulation.

**Hepatic Portal Vein** - This vein is formed by the union of two large veins and many smaller tributaries. The two larger veins are the:

1. **Superior Mesenteric** - This is the larger of the two veins. It extends further caudally and drains a greater area. Its tributaries are the:
   - **Posterior Pancreaticoduodenal** - from the duodenum and pancreas.
   - **Inferior Mesenteric** - from the large intestine
   - **Ileocolic** - from the cecum, adjacent ileum, and colon
   - **Intestinal** - numerous brances arising from the small intestines
2. **Gastrosplenic** - This vein comes from the stomach, spleen, and the left edge of the pancreas. Its tributaries include the:
   - **Anterior Splenic** - from the cranial end of the spleen
   - **Posterior Splenic** - from the caudal end of the spleen
   - **Left Gastroepiploic** - from stomach and greater omentum
   - **Middle Gastroepiploic** - from stomach and greater omentum
   - **Pancreatic** - from the pancreas

Above the union of these two large veins, several smaller veins join the hepatic portal vein. These are the:

**Coronary** - from the lesser curvature of the stomach
**Anterior Pancreaticoduodenal** - from the duodenum and the pancreas
**Right Gastroepiploic** - from the greater curvature of the stomach

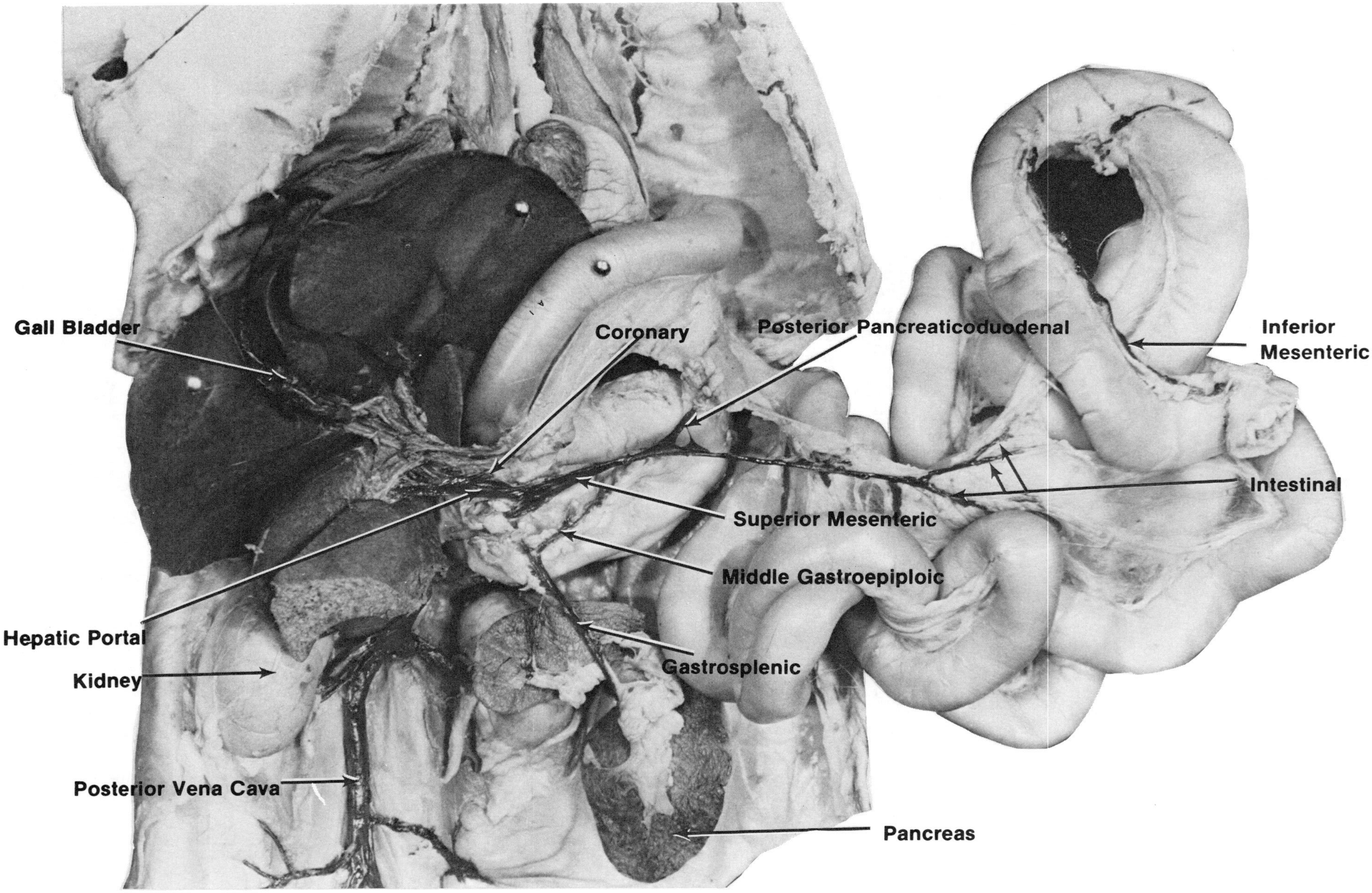

VEINS - HEPATIC PORTAL SYSTEM

# ARTERIES - ANTERIOR REGION

Remove the major veins surrounding the heart. This will enable you to expose the arteries. Use your dissecting needle to clear arteries of connective tissue, to separate them, and to follow them.

**Pulmonary Artery** - On the ventral surface of the heart passing dorsally and to the left is the large *pulmonary artery.* It originates in the right ventricle. Try tracing its two branches to the lungs. The pulmonary is the only artery injected with blue latex to indicate that it carries deoxygenated blood. All other arteries will appear red or pink because they have been injected with red latex to indicate that the blood transported in them is oxygenated.

**Aorta** - Locate the *aorta,* the largest systemic artery of the body. It leaves the left ventricle and passes to the left, dorsal to the pulmonary artery.

Near its origin the aorta gives off two branches, the *right* and *left coronary arteries.* These pass over the surface of the ventricles sending branches deep into the myocardium, supplying the heart muscle with necessary nutrients and oxygen.

From its origin in the left ventricle the aorta curves to the left, passes dorsally and continues downward along the left side of the vertebral column. The proximal curved portion is called the *aortic arch* while the next segment of the aorta within the thorax is known as the *thoracic aorta.*

The aortic arch gives rise to arteries that supply the neck, head, shoulders, and forelimbs. Whereas in man three arterial trunks arise from the aortic arch, in the cat there are only two. These are the:

1. **Brachiocephalic** (or Innominate) - This is the first arterial branch of the aorta. As its name indicates it supplies blood to the forelimb and head. At the level of the second rib it divides into the:

   **Right Subclavian** - This artery leaves the thorax, gives off four major branches, then continues as the *axillary artery* in the armpit and shoulder region, then as the *brachial artery* in the upper forelimb. In the lower forelimb it branches to form the *radial* and the *ulnar arteries.*

   **Right and Left Common Carotids** - These two branches of the brachiocephalic artery rise cranially and lie parallel to each other along either side of the trachea.

2. **Left Subclavian** - The second branch of the aortic arch is the *left subclavian artery.* It gives off branches similar to the right subclavian, then proceeds into the left forelimb as the *left axillary*, the *left brachial,* the *left radial,* and the *left ulnar* arteries.

Near the first rib the *subclavian artery* gives off four branches. These are the:

**Internal Mammary** - This artery passes caudally to the ventral thoracic wall and supplies the adjacent muscles, pericardium, and mammary glands with blood.

**Vertebral** - This artery passes cranially through the vertebral canal of the cervical vertebrae giving off branches to deep neck muscles and to the spinal cord. It enters the foramen magnum of the skull to form the *basilar artery* which supplies the hind brain.

**Costocervical** - This artery arises from the dorsal surface of the subclavian supplying the muscles of the neck and chest. At the upper border of the scapula it is called the *transverse scapular artery* and it supplies the shoulder muscles.

The *common carotid arteries* in their passage cranially give off the following arteries:

**Superior Thyroid** - This artery supplies the thyroid gland, laryngeal muscles, ventral neck muscles.

**Internal Carotid** - This artery enters the skull and joins the *posterior cerebral artery* which supplies the brain.

**External Carotid** - This artery is a continuation of the common carotid artery after it has given off the internal carotid artery.

**Lingual** - This artery is the major branch of the external carotid artery. It gives off branches ot the hyoid, pharyngeal muscles, then enters the tongue.

**External Maxillary** - This artery is a branch of the external carotid near the angle of the jaw. It passes along the ventral border of the masseter muscle and gives off branches to the lips and mouth.

**Posterior Auricular** and **Superficial Temporal** - These two arteries are distal branches of the external carotid artery running dorsally and laterally to supply the muscles of the side and back of the head. The *posterior auricular artery* is located near the ear, the *superficial temporal artery* is found near the temple.

Some of the branches of the *axillary artery* are the:

**Ventral Thoracic** - This artery passes caudally from the axillary area to supply the pectoral muscles near the sternum.

**Long Thoracic** - It too passes caudally to the pectoral muscles and to the latissimus dorsi.

**Subscapular** - This artery is the largest branch of the axillary artery. It passes laterally and dorsally to supply the dorsal shoulder muscles. It then branches to supply the teres major, latissimus dorsi as well as the triceps muscles.

Trace as many of these arteries in the anterior region as time allows. You may come to the lab during your free time to continue these dissections.

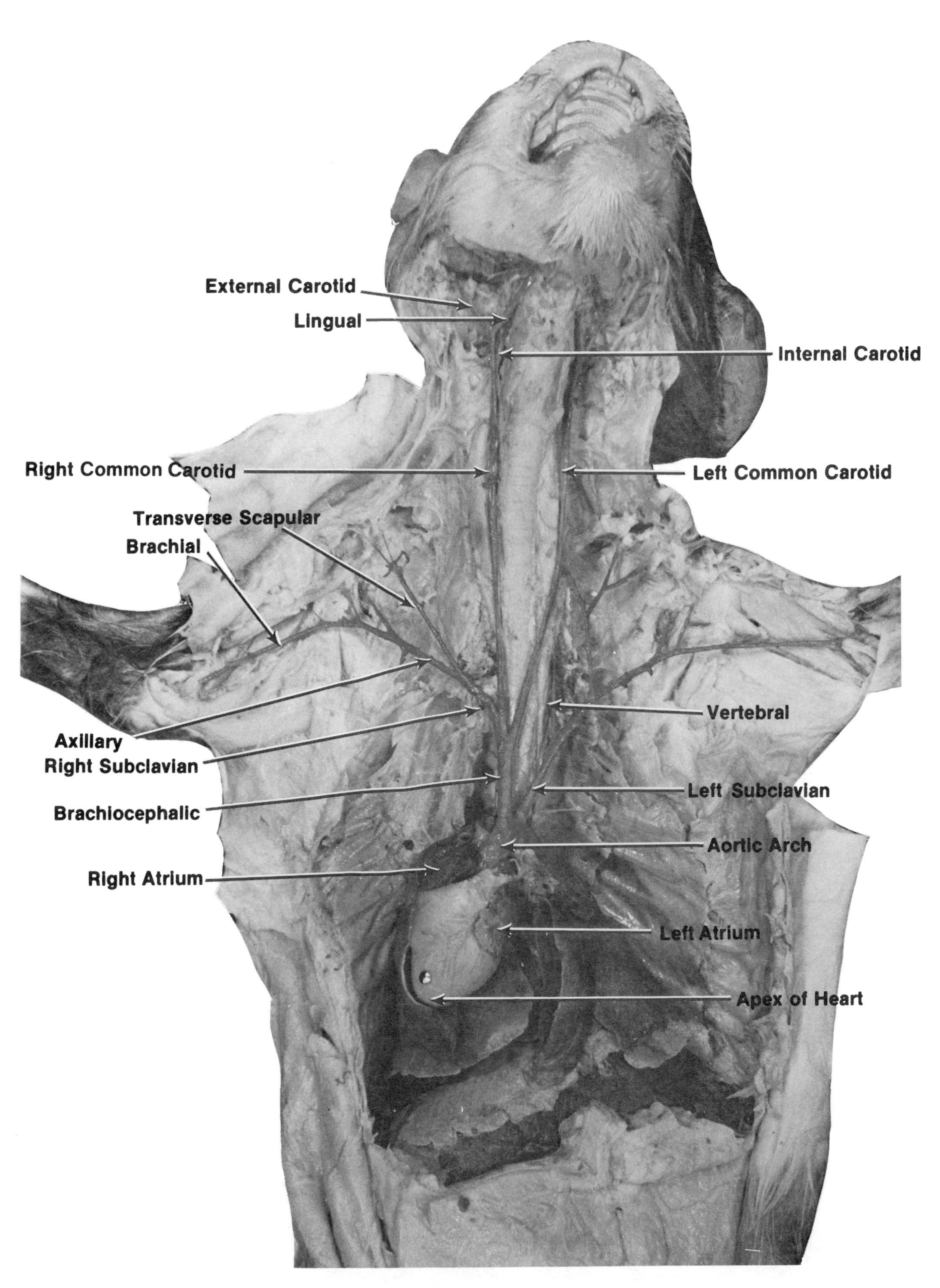


ARTERIES - ANTERIOR REGION

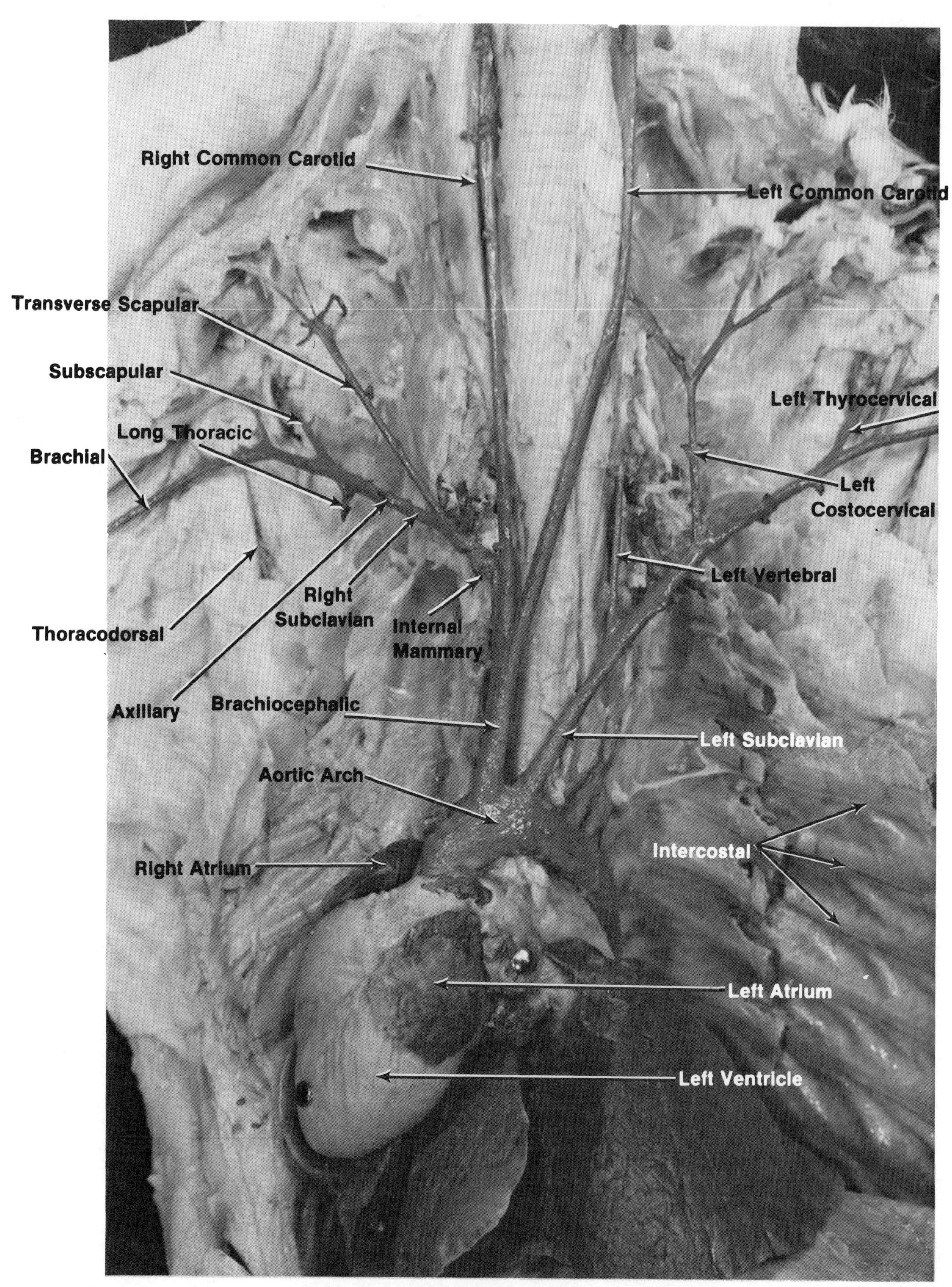


ARTERIES - ANTERIOR REGION (CLOSE-UP)

# ARTERIES - POSTERIOR REGION

Follow the *aortic arch* dorsally and caudally through the thorax. Push the left lung toward the right and observe the *thoracic aorta* along the dorsal body wall to the left of the vertebral column. In the photo the lungs have been removed to faciliate observation.

**Intercostal Arteries** - Between each two ribs note the *intercostal arteries.* They are given off by the *thoracic aorta* to the right and left sides to supply the intercostal muscles.

Other branches of the thoracic aorta above the diaphragm include the:

**Bronchial** - These arteries supply the lung tissues with oxygenated blood.
**Esophageal** - This artery supplies the esophagus.
**Phrenic** - This artery supplies blood to the diaphragm. It arises from the dorsal aorta in the vicinity of the diaphragm, at times anterior, at times posterior to the diaphragm.

The aorta contines caudally and passes through the diaphragm into the abdominal cavity. It is here known as the *abdominal aorta.*

**Celiac** - Immediately posterior to the diaphragm the *abdominal aorta* gives off an unpaired vessel, the *celiac artery,* which supplies blood to the organs of the upper abdomen.

The three branches of the *celiac artery* are the:

1. **Hepatic** - This artery is located in the lesser omentum together with the common bile duct and hepatic portal vein. It passes under the stomach to the liver. It gives off some branches to other organs as well:
**Cystic** - This artery supplies the gall bladder, then sends branches to the liver.
**Gastroduodenal** - This largest branch of the hepatic artery begins near the pyloric portion of the stomach. It, in turn, gives rise to three branches:
**Pyloric** - to the pylorus and lesser curvature of the stomach
**Anterior Pancreatoduodenal** - to the duodenum and pancreas
**Right Gastroepiploic** - to the greater omentum and the greater curvature of the stomach.
2. **Left Gastric** - This is the second branch of the *celiac artery.* It lies along the lesser curvature of the stomach. It anastomoses with the pyloric artery.
3. **Splenic** - This is the third and largest branch of the *celiac artery.* It gives off branches to the stomach, spleen, and pancreas.

**Superior Mesenteric** - This artery is the next branch of the abdominal aorta below the celiac artery. It supplies the small intestine and parts of the large intestine. Its branches are the:
**Posterior Pancreaticoduodenal** - to the caudal portion of the pancreas and duodenum
**Middle Colic** - to the ascending colon

**Ileocolic** - to the cecum and ileum
**Intestinal** - Numerous branches of this artery pass through the mesentery to supply blood to the small intestines.

Other branches of the *abdominal aorta* supplying the abdominal visceral are the:

**Inferior Mesenteric** - This unpaired artery passes to the descending colon and rectum. It has the following branches:
**Left Colic** - to the descending colon
**Superior Hemorrhoidal** - to the rectum

There is extensive anastomosing between the arteries in the areas of the large and small intestines.

The *abdominal aorta* continues to pass cuadally and give off the following arteries:

**Adrenolumbar** - paired vessels to the adrenal glands and lumbar body wall
**Renal** - paired arteries to the kidneys
**Genital (Internal Spermatic** or **Ovarian)** - In the male these paired arteries pass through the inguinal canal to the *testes.* In the female they supply the *ovaries* and *uterus.*

The abdominal arteries studied thus far are primarily those that supply blood to the abdominal viscera. Several *parietal* vessels supplying the body wall also branch from the aorta. These include the:

**Lumbar** - several paired arteries supplying the body wall
**Iliolumbar** - These are also paired vessels to the body wall.
**External Iliac** - At the level of the last lumbar vertebra the *abdominal aorta* divides into two terminal branches which pass to the hind limbs. These are the *external iliac arteries.* Note that there is no common iliac artery as in man, instead, the paried *internal iliac (hypogastric) arteries* also arise independently from the aorta.
**Deep Femoral** - The external iliac artery gives off this vessel before passing into the thigh.
**Femoral** - Within the thigh the external iliac continues as the *femoral artery.* It gives off the following branches:
**Saphenous** - The continuation of the femoral artery beyond the thigh. It lies upon the medial surface of the hind leg along with the *greater saphernous vein* and the *saphenous nerve.* Several large branches to the leg musculature are also given off. They include the:
**Superior Articular** - This artery passes laterally to the knee.
**Popliteal** - This is a continuation of the femoral artery into the popliteal fossa. It then continues and divides to form the:
**Anterior Tibal** - This artery passes ventrally through the leg into the foot.
**Posterior Tibial** - This artery terminates in the dorsal crural muscles of the cat. In humans it continues into the foot.

Return to the *deep femoral artery.* Near its origin it gives off several branches. These include the:

**Inferior Epigastric** - This artery passes anteriorly along the inner surface of the rectus abdominis muscle to anastomose with the *internal mammary artery.*
**Vesicular** - to the bladder
**External Spermatic** - to the spermatic cord and the testes

Return to the *internal iliac arteries.* They arise from the dorsal aorta immediately posterior to the base of the external iliac arteries. Several branches are given off but these are rarely well injected and will therefore be difficult to trace. These include the:

**Umbilical** - This artery is the largest branch of the internal iliac. It supplies the urinary bladder. It is the remnant of the *allantoic* or *umbilical artery* of the embryonic circulation.

**Superior Gluteal** - supplies several muscles of the thigh

**Middle Hemorrhoidal** - This artery passes to the lateral wall of the rectum and urethra. In females the *uterine artery* extends to the uterus and its horns.

**Inferior Gluteal** - This artery passes to the gluteus medius and other muscles in the thigh.

**Median Sacral** and **Caudal** - The aorta continues caudally beyond the iliac branching. It is, however, a very small vessel. It passes down the sacrum as the *median sacral artery* and into the tail as the *caudal artery*.

Locate these arteries in your specimen.

**Label the parts of the human arterial system below and compare them with those you observed in the cat.**

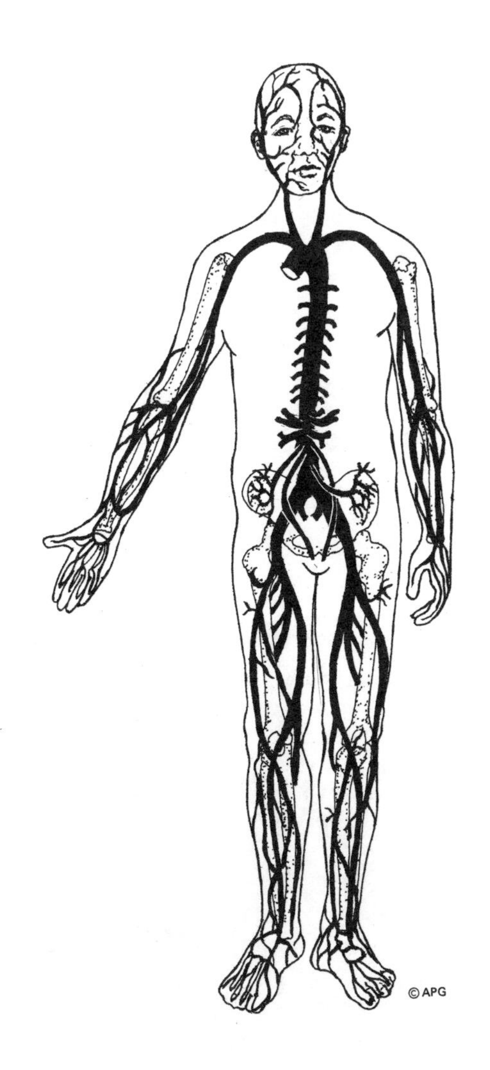

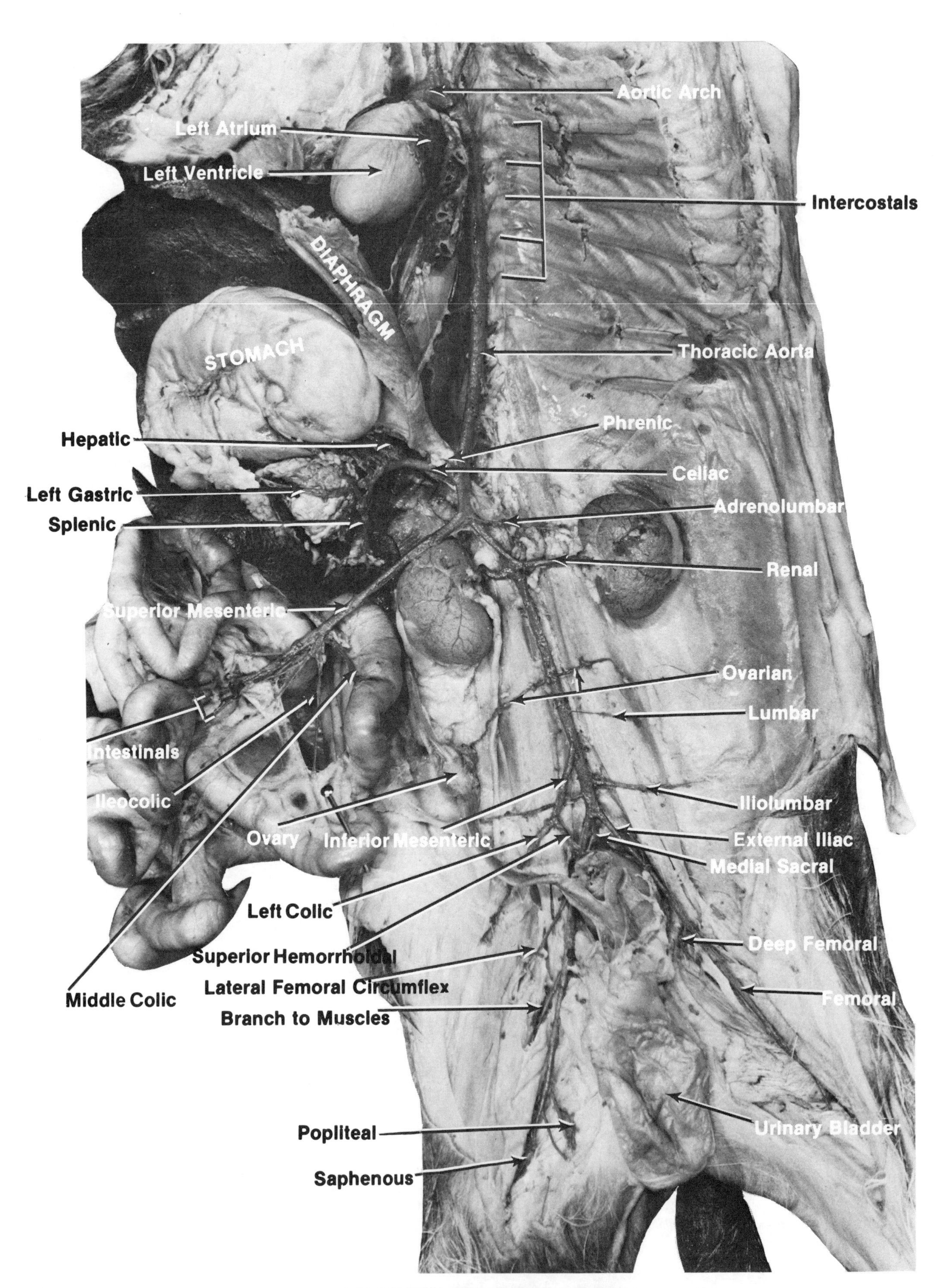


ARTERIES - POSTERIOR REGION

Name. . . . . . . . . . . . . . . . . . . . . . . Section . . . . . . Date. . . . . . . . . . . . . . . . . . . .

# SELF - QUIZ VI
# VEINS AND ARTERIES

1. Name the large veins and arteries that carry blood to and from the kidneys.
2. The jugular veins drain the head of blood. Name the large arteries lying alongside the jugulars that carry blood to the head.
3. Name the three major blood vessels that join to from the hepatic portal vein.
4. Name the blood vessels supplying the heart muscle with blood.
5. Name the major branches of the brachiocephalic artery.
6. Name the major branches of the abdominal aorta.
7. Name the genital arteries of male and female cats.
8. Name the major branches of the external iliac arteries.
9. Name the major veins that empty into the innominate veins.
10. For each of the veins named below, tell from what parts of the body they carry blood.

ANSWERS

1. ____________________
2. ____________________
3. ____________________
4. ____________________
5. ____________________
6. ____________________
7. ____________________
8. ____________________
9. ____________________
10. a. brachiocephalic ____________________
    b. femoral ____________________
    c. intercostals ____________________
    d. greater saphenous ____________________
    e. gastrosplenic ____________________
    f. mesenteric ____________________
    g. brachial ____________________
    h. cephalic ____________________
    i. radial ____________________
    j. phrenic ____________________

Label all of the features indicated on the photograph.

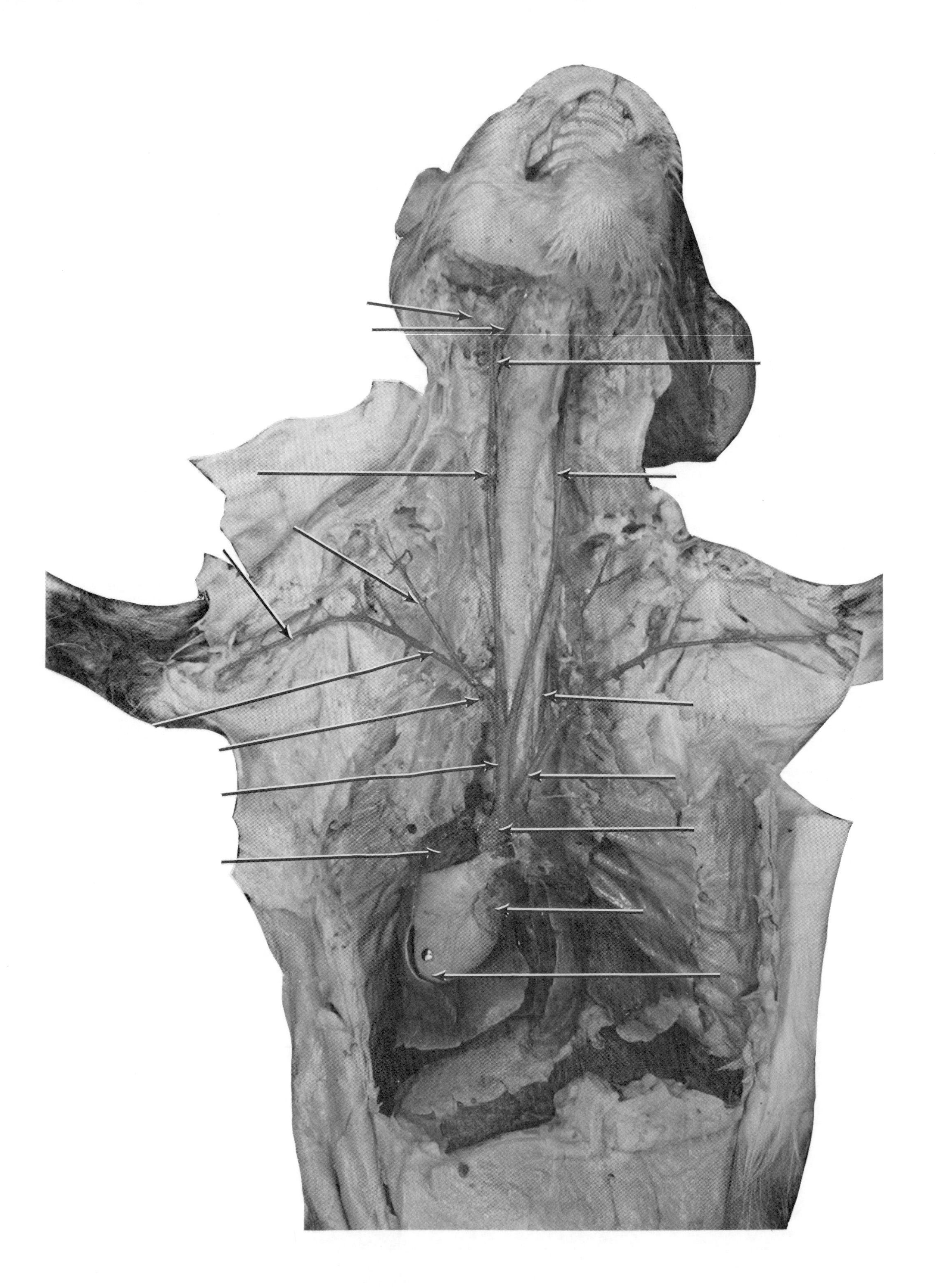

# THE UROGENITAL SYSTEM - FEMALE

Remove the liver, spleen, stomach, and intestines. Leave the last two inches of the large intestine intact.

## Urinary System

**Kidneys** – They are large bean-shaped structures on either side of the vertebral column at the level of the third to fifth lumbar vertebrae. Although they bulge into the abdominal cavity, they lie beneath the peritoneum, or *retroperitoneally,* often surrounded by fat. The *adrenal glands* are narrow band-like structures lying median to the anterior region of the kidneys.

Clear the kidneys to expose the *renal arteries, renal veins,* and the *ureters*. Some of the parts of the kidney are the:

**Hilus** – This is a central depression in the medial surface of the kidney. The ureters exit the kidney at the hilus.

In order to observe the following structures it is necessary to cut one of the kidneys in frontal section as in the diagram below.

**Renal Sinus** – This is a central cavity which contains fat, branches of the renal vessels and the *renal pelvis.* The pelvis is the funnel-shaped expanded portion of the ureter within the renal sinus.

**Renal Cortex** – This is the outer layer of kidney tissue.

**Renal Medulla** – This is the more central portion of the kidney, beneath the cortex.

**Renal Papilla** – This is a cone-shaped projection of the medulla enclosed by the pelvis. In the cat's kidney there is only one papilla, in man there are many.

Trace the ureter from the hilus to the *urinary bladder.* Do this on both sides. Do not injure the reproductive structures. Lift the urinary bladder and find the *urethra*, which transports urine from the bladder.

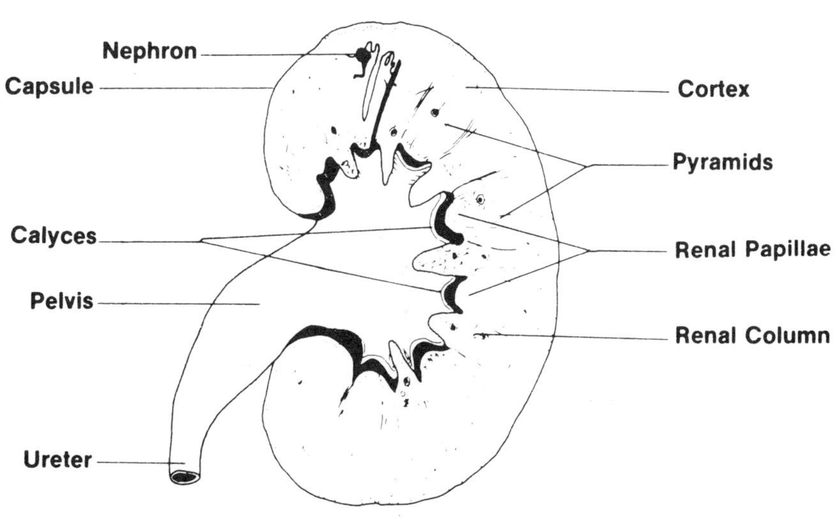


**HUMAN KIDNEY (Frontal Section)**

To this point only the urinary structures have been examined. They are alike in males and females.

If your specimen is a female, continue the dissection as directed here. If your specimen is male, continue in the next section. However, whether your specimen is male or female, you are responsible for knowing the reproductive structures of each. Therefore, work closely with a student whose cat is of the opposite sex of your specimen.

## Genital System

**Ovaries** — These are the female gonads. They are paired, small bean-shaped structures located posterior to the kidneys. The *oviducts*, or fallopian tubes, and extremely narrow, usually located dorsal to the anterior portion of the ovary. Use a hand lens to observe them more closely. Also observe the expanded ends of the openings, the *ostium*, fringed by small finger-like projections, termed *fimbriae*. These guide the ova into the ostium.

**Uterine Horns ( Cornua)** — Trace the oviducts to the dorsal surface of the ovary where they join the much wider *uterine horns*. Trace these caudally to where they join to form the body of the *uterus*, which lies dorsal to the urinary bladder and urethra.

In cats and other mammals, the fetus does not develop in the body of the uterus, as in man, but in the horns extending from the uterus. This permits the development of more fetuses at one time and the birth of a litter. In humans, development of the fetus in the body of the uterus makes multiple births a rarity.

**Membranes** — The ovaries are suspended from the dorsal body wall by a peritoneal membrane called the *mesovarium*. An *ovarian ligament* connects the ovaries to the uterine horns. Each horn is supported by a peritoneal fold the *mesometrium*. These three membranous suspensions are part of the *broad ligament*. This ligament extends into the pelvic area serving to hold the body of the uterus and vagina to the body wall. Another support, the *round ligament*, extends from the dorsal body wall to the middle of each uterine horn.

In order to continue the dissection it is necessary to cut through the *pubic symphasis* and spread the pelvic bones apart. This will expose the rest of the genital organs lying below the pelvic bones.

**Uterus and Vagina** — You are now ready to expose the entire urethra and the body of the *uterus*. The *vagina* is a continuation of the uterus and lies dorsal to the urethra. The *cervix* is a constricted area between the body of the uterus and the vagina.

**Urogenital Sinus** — Separate the urethra from the vagina. Posteriorly the vagina and urethra unite to form a common passageway which opens to the exterior, the *urogenital sinus*, or *vestibule*. In human females the vagina and urethra are separate throughout their lengths and the vestibule is a much reduced area, a part of the external genitalia.

**External Genitalia** — Follow the urogenital sinus caudally to its opening on the outside of the body. Use your scissors to make a ventral longitudinal cut along the urogenital sinus in an anterior direction from the external genitalia, through the vagina beyond its union with the urethra, as in the accompanying photo. Find the *urethral orifice* on the ventral surface of the vagina at its junction with the urethra. Near the opening of the urogenital sinus, again in the ventral wall, locate the *clitoris*, the homolog of the penis. On either side of the *urogenital aperture*, at the external opening of the urogenital sinus, just ventral to the anus, are folds of skin called the *labia majora*. These, together with the urogenital aperture, constitute the *vulva*.

**The Pregnant Cat** - The last photo is this section shows the reprodutuve organs of the cat during pregnancy. The gestation period for cats is about 60 days. The cat in the photo is about at midterm. Note the following:

**Loculi** - These are the enlargements along the horns of the uterus of the pregnant cat. By counting these, the number of fetuses may be determined. There are usually four to six in a litter. In the photo we see two such loculi, however, the uterine horns on the right side, which had also contained two fetuses were removed. The size of each loculus is determined by the stage of development of the fetus and its *placenta*.

**Zonary placenta -** Open one of the loculi to observe the *serous* outer layer, the muscular middle layer, and the inner *endometrium.* This last layer is thin except in the area of *zonary placenta* which is a broad band surrounding the center of the fetus.

Note the *uterine blood vessels* that surround and drain the loculi. As seen in the photo they are almost as thick as the posterior vena cava. Compare this to the extrememly narrow *ovarian* artery seen in the photo on page 120 where the non-pregnant female cat is seen.

## HUMAN FEMALE - REPRODUCTIVE STRUCTURES

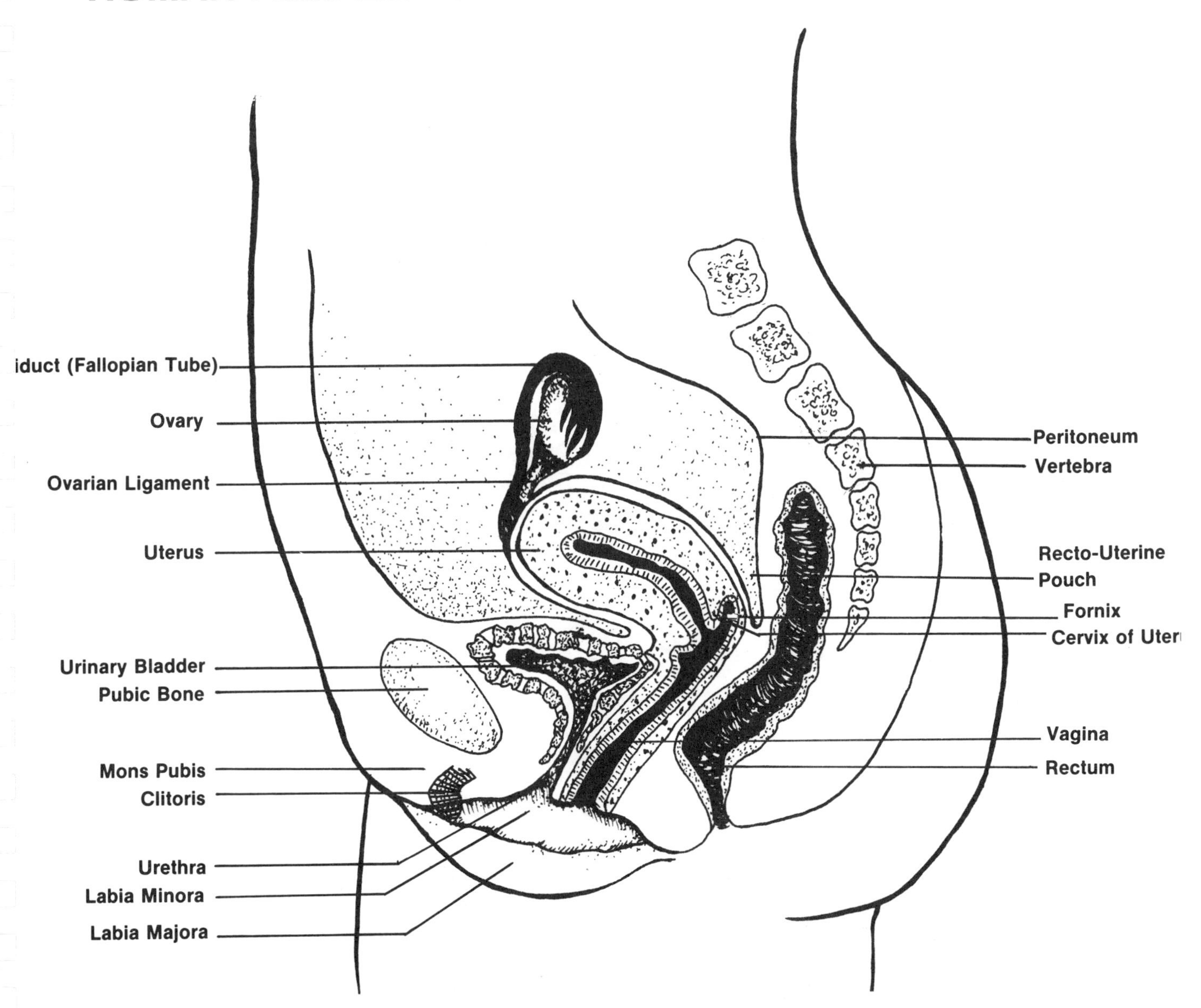

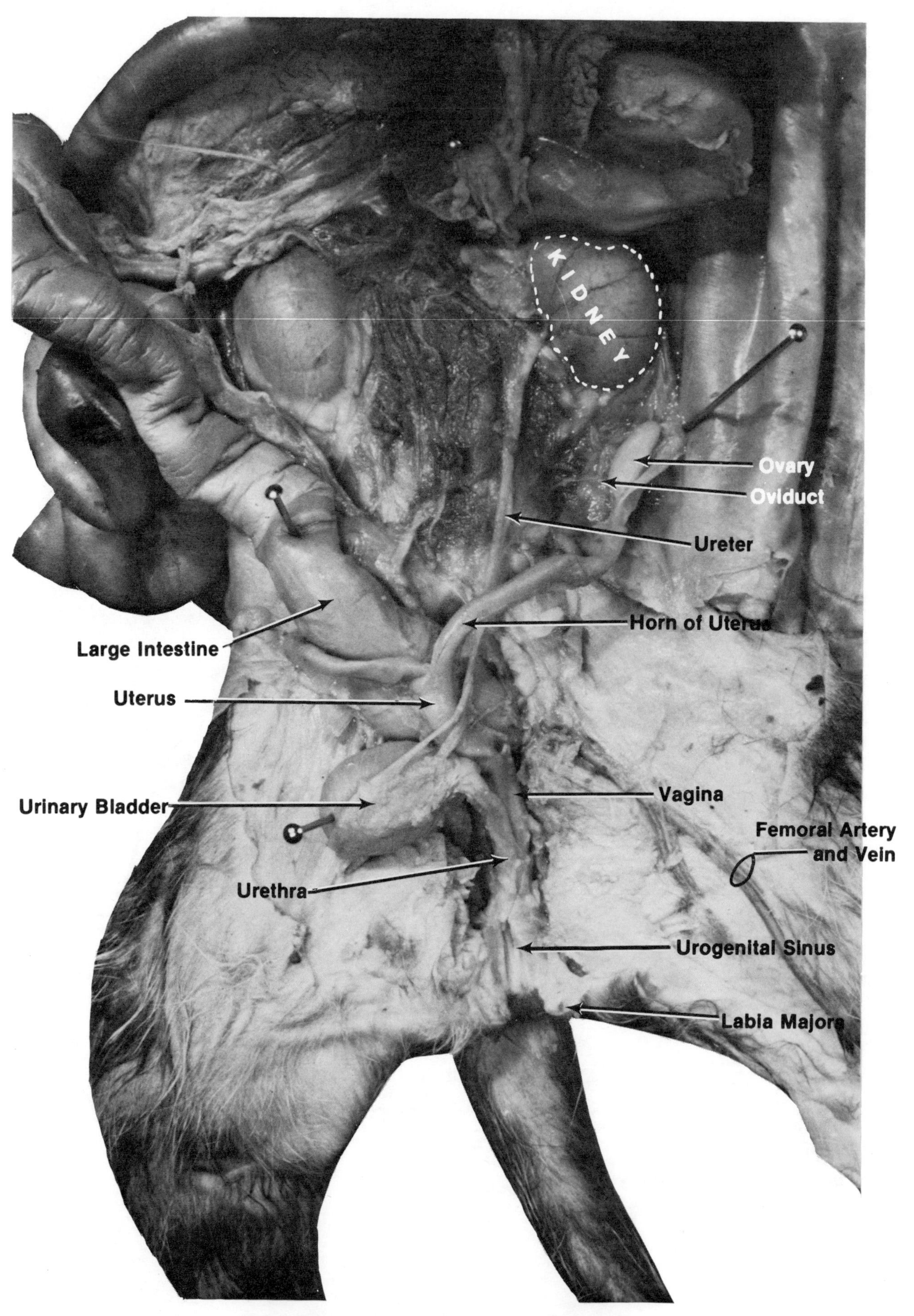


THE UROGENITAL SYSTEM - FEMALE

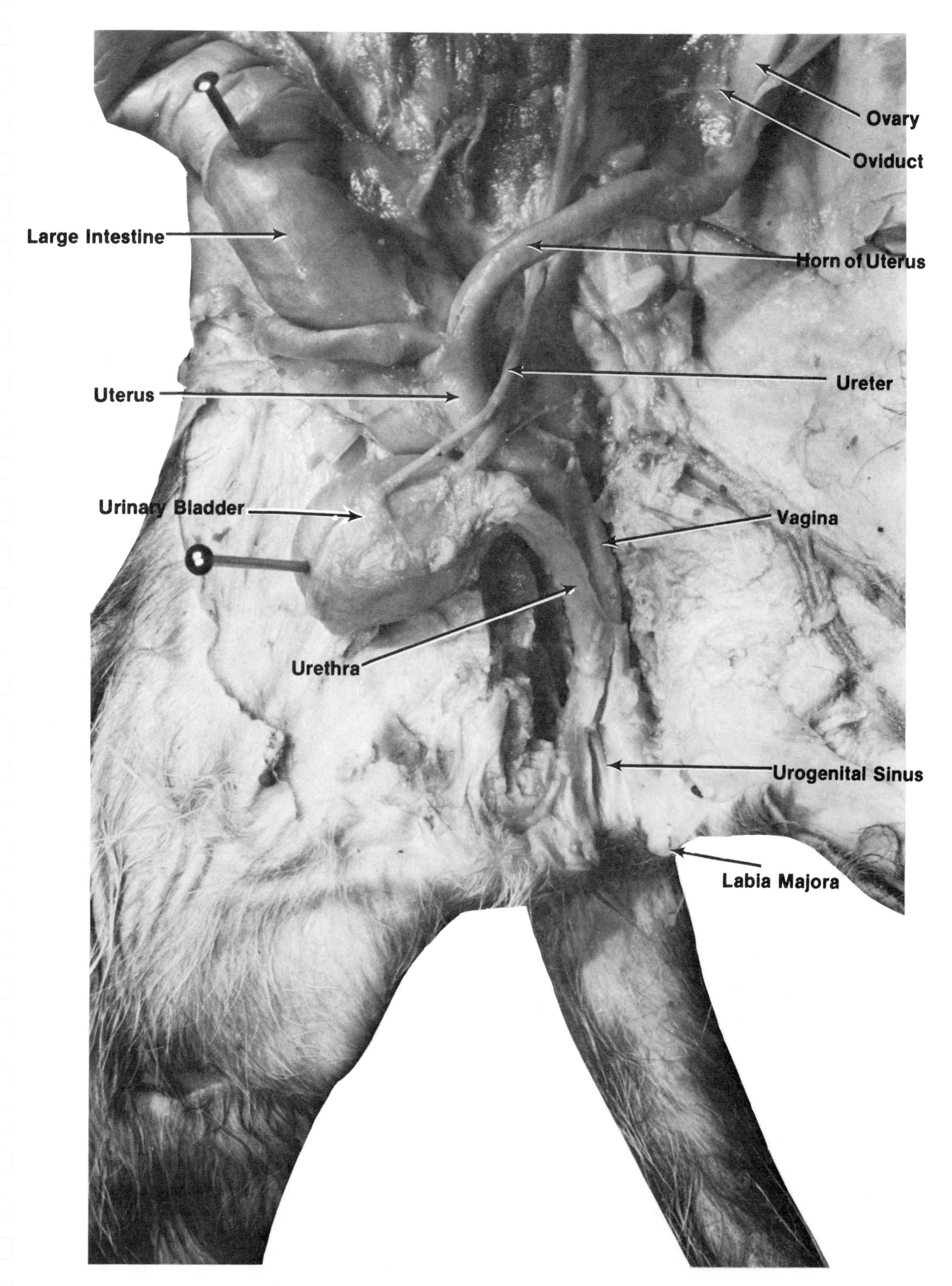


THE UROGENITAL SYSTEM - FEMALE (CLOSE-UP)

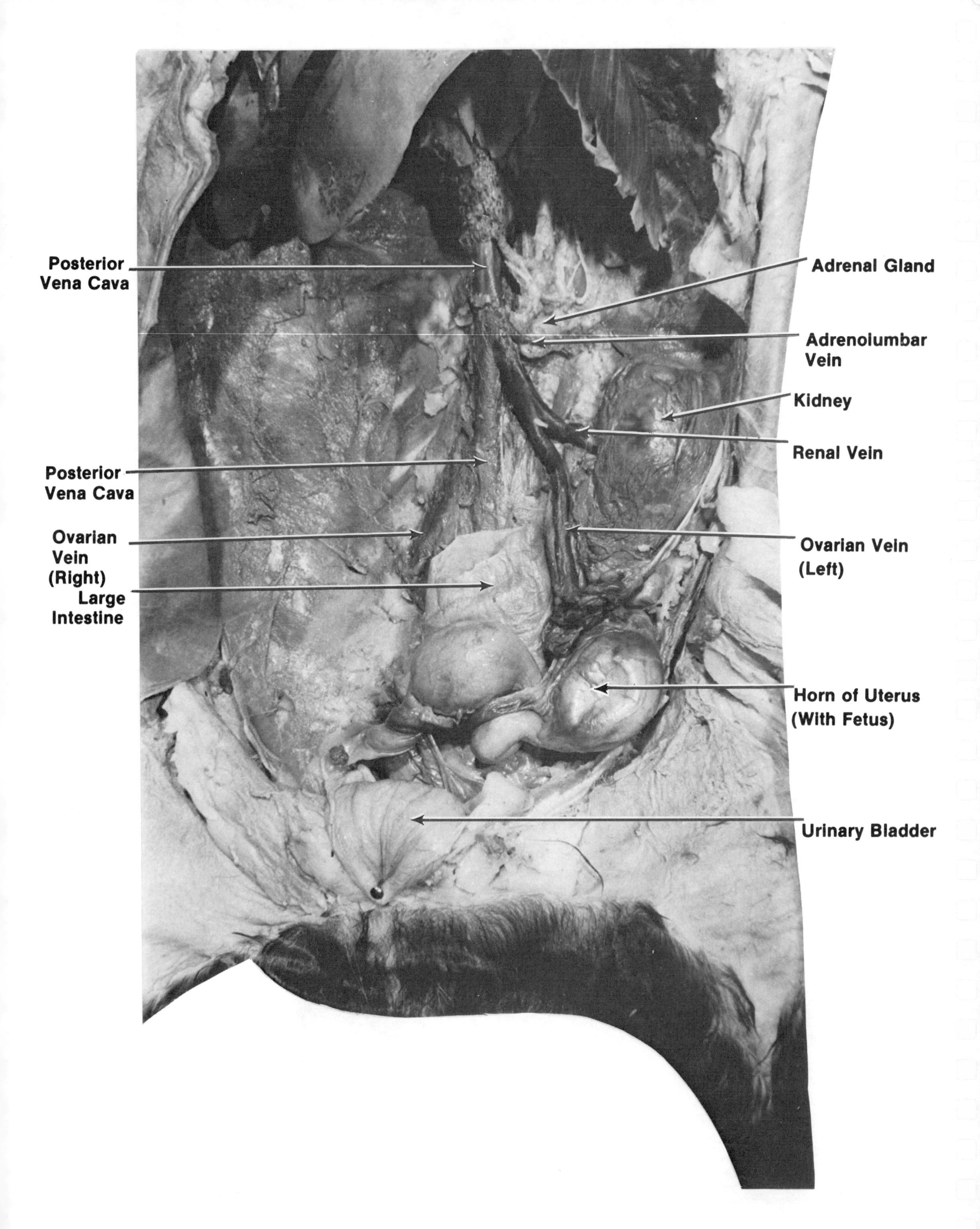


THE UROGENITAL SYSTEM - PREGNANT FEMALE

# THE UROGENITAL SYSTEM - MALE

The urinary organs of male cats, the *kidneys, ureters, urinary bladder*, and *urethra*, are similar to those of the female. Since these have already been described in the last section they will not be repeated here.

The male reproductive structures include the:

**Testes** - These are the male gonads. Locate the *scrotum*, the swollen double sac ventral to the anus. They contain the *testes*. Carefully cut the skin of the scrotum. It is lined with peritoneum and divided into two compartments by a median septum. In the photo the left testis is shown in place, the right testis was placed upon the right thigh to facilitate observation.

The *testes* are oval structures in a membrane bag, the *fascial sac*. The covering membranes are the *tunica vaginalis* and a tougher fibrous coat, the *tunica albuginea*. The posterior end of the testis is attached to the membranous sac within the scrotum by a ligament, the *gubernaculum*, which is homologous to the round ligament of the ovary.

**Epididymis** - This is an extensively coiled tubular scructure lying on the dorsal surface of each testis. It consists of a *head* on the anterior part of the testis where it is connected to the testis by numerous microscopic efferent ductules. It also has a *body*, the middle portion, and a posterior portion, the *tail*. Follow the epididymis tail cranially. Its convoluted ducts are continuous with the duct that exits the scrotum into the abdominal cavity.

**Ductus Deferens (Vas Deferens)** - It is through this tube that sperm and seminal fluid leave the testes. It exits the scrotum into the abdominal cavity.

Note: **In order to study the remainder of the male reproductive system cut the pelvic bone at the pubic symphasis and spread the pelvis apart.**

**Spermatic Cord** - The ductus deferens is only one of the tubes leaving the testis. Blood vessels, nerves, and lymphatic vessels supplying the testis also pass from the scrotum into the abdominal cavity. They are united by a tough outer fascia to form the *spermatic cord*.

**Inguinal Canal** - Follow the spermatic cord cranially through the *external inguinal ring* located at the juncture of the scrotum and the abdominal wall. Continue further cranially through a short **channel in the abdominal wall, the *inguinal canal*,** and out into the abdominal cavity through the *internal inguinal ring*.

During embryological development the testes are at first located within the abdominal cavity, below the kidneys. During later development they descend through the *inguinal canal* into the scrotum. In human males the condition of *inguinal hernia* is common. It is a weakening of the inguinal rings permitting a loop of the intestine to be pushed through the inguinal canal into the scrotum. This condition is due to man's upright, two-legged position. Cats do not suffer from this malady.

**Upper Spermatic Cord** - Continue to follow the spermatic cord within the abdomen. The blood vessels, the *spermatic vein* and *internal spermatic artery* soon separate from the ductus deferens. As seen in the photo the *right spermatic vein* enters the posterior vena cava below the level of the kidney, while the *left spermatic vein* enters the renal vein near the top of the kidney. It thus does not enter the vena cava directly. The *right and left spermatic arteries* enter the abdominal aorta next to one another at the level of the base of the kidney. This can be seen clearly in the photo. Trace these blood vessels on your specimen.

**Upper Ductus Deferens** - The ductus deferens loops dorsally over the base of the *ureter* near the urinary bladder and continues caudally. The *urethra* emerging from the urinary bladder, together with the ductus deferens, pass posteriorly and penetrate the *prostate gland* at the proximal end of the *penis.* From that point on, the urethra continues as a merged tube, the *urogenital canal* carrying sperm and seminal emissions from the testes and prostate gland, plus urine from the urinary bladder. There are no seminal vesicles or ejaculatory duct in the cat.

**Penis** - Follow the urethra, or urogenital canal, caudally to the beginning of the *penis.* The penis is the cylindrical copulatory organ of males. Remove the skin and trace the penis to its attachment in the region of the pubic symphasis. Locate the *crus* of the penis and a small muscle the *ischiocavernosum.* Both are lateral projections at the proximal end of the penis anchored to the ischium bone. The crus is the proximal end of the *corpus cavernosum,* a cylindrical mass of vascular erectile tissue which, together with a second corpus cavernosum lying side by side, form the dorsal part of the penis. A third cylindrical mass of vascular erectile tissue lies ventrally within the penis in a groove between the corpora cavernosum. This is the *corpus spongiosum.* The urethra passes through this mass of vascular tissue.

Cut the *crus* of the penis and the *ischiocavernosum muscle* from their attachments on the ischium. Locate the *bulbourethral glands (Cowper's glands)* dorsal to these structures.

Locate the distal end of the penis. Its terminal portion is a conical structure known as the *glans penis.* Here the urethra opens to the outside, the opening known as the *urogenital aperture.* A sheath of skin called the *foreskin,* or *prepuce,* covers the glans penis. During circumcision in humans the foreskin is removed. In the photo a small portion of foreskin remains attached to the glans in order to indicate its normal position, the remainder was removed to expose the glans. Minute horny papillae cover parts of the glans.

Each student is responsible for learning the reproductive structures of both male and female cats.

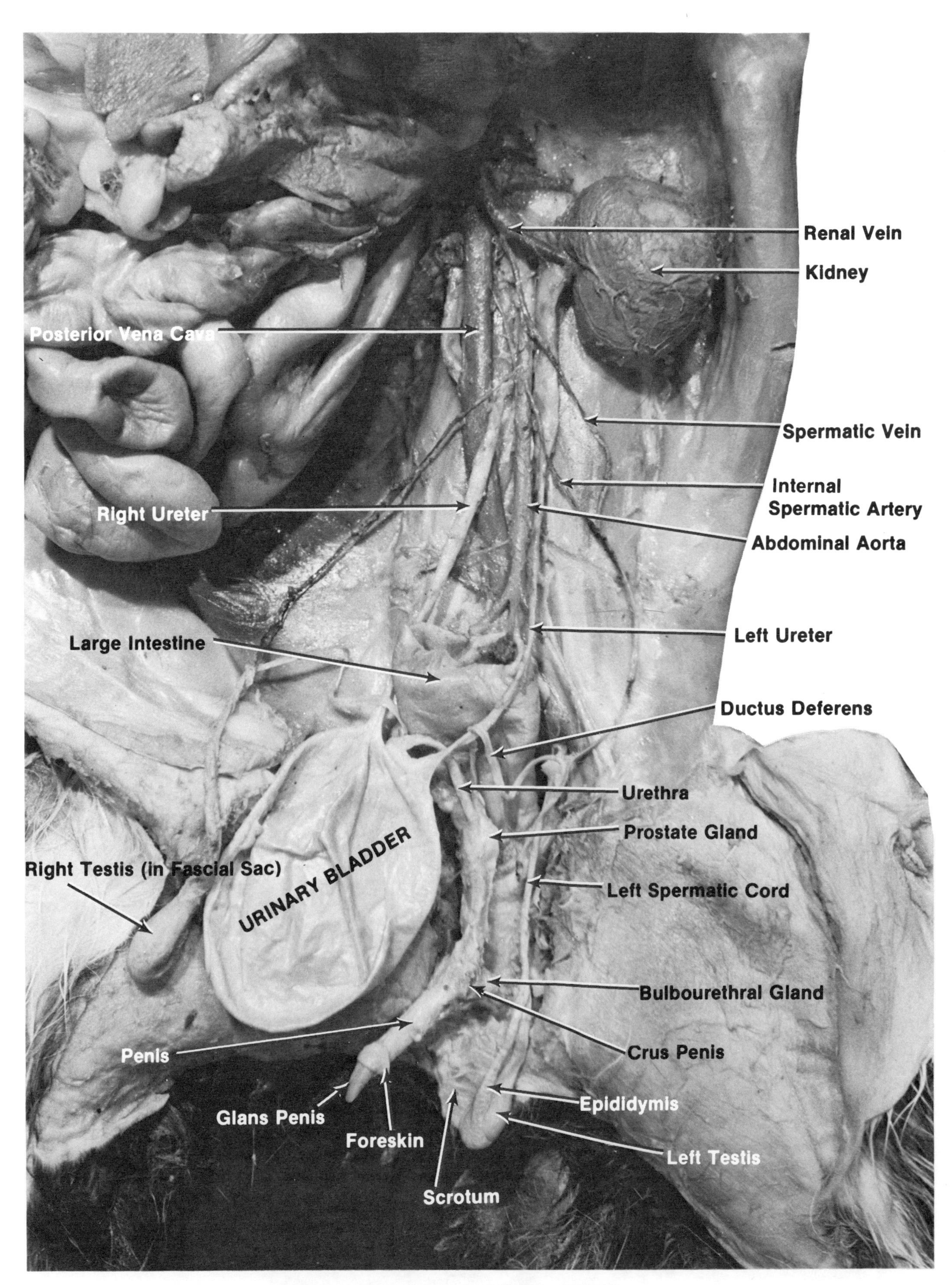


THE UROGENITAL SYSTEM - MALE

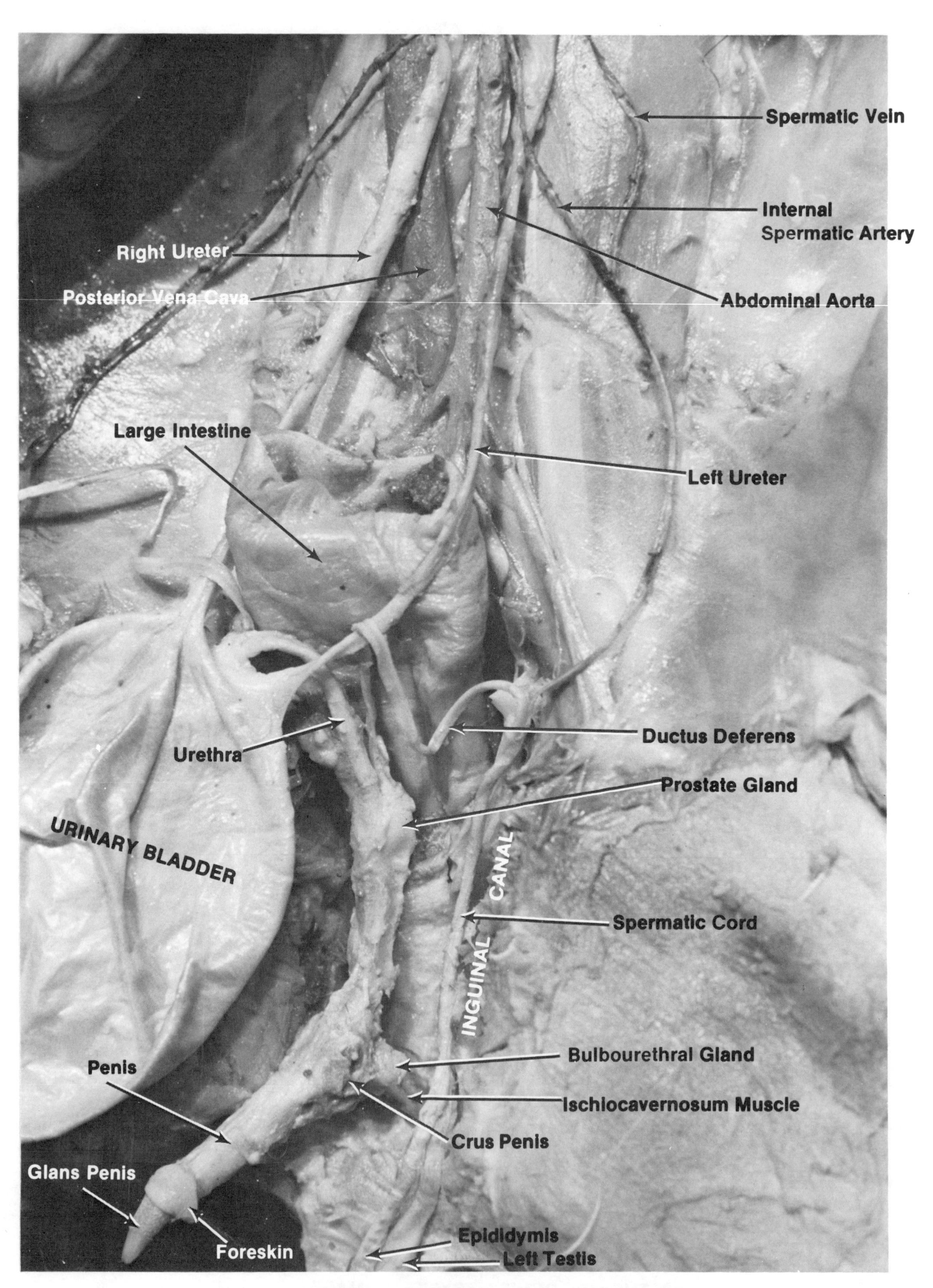


THE UROGENITAL SYSTEM - MALE (CLOSE-UP)

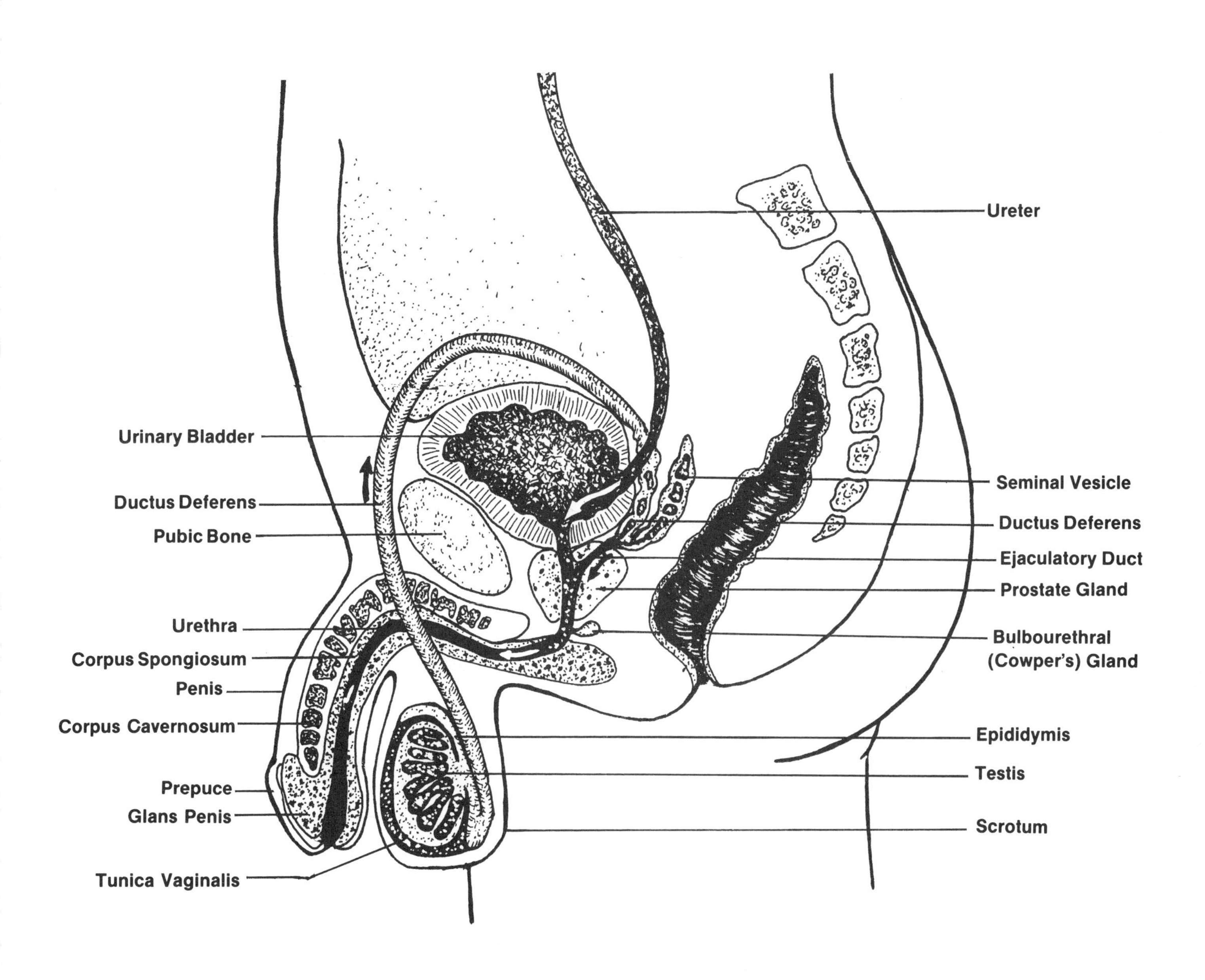


# HUMAN MALE - REPRODUCTIVE STRUCTURES

Name ................................ Section ....... Date ................

# SELF - QUIZ VII
# UROGENITAL SYSTEM

1. The human fetus develops in its mother's uterus. In what structure does the cat embryo develop?
2. The vagina and urethra join to form a single duct in the, a) female cat, b) human female, c) both, d) neither.
3. The urethra and vas deferens join and exit as a single duct in, a) male cat, b) human male, c) both, d) neither.
4. Name the tubes that carry urine from the kidneys to the urinary bladder.
5. Name the blood vessels supplying blood to the testes and those that drain the testes.
6. Name the several membranous coverings of the testes.
7. What vessels and structures comprise the spermatic cord?
8. Name the duct that carries urine from the urinary bladder to the outside of the body.
9. Besides the testes and penis, name four accessory male reproductive structures.
10. Define each of the terms listed below.

ANSWERS

1. ____________________
2. ____________________
3. ____________________
4. ____________________
5. ____________________
6. ____________________
7. ____________________
8. ____________________
9. ____________________
10. a. fallopian tubes ____________________
    b. gubernaculum ____________________
    c. prostate ____________________
    d. Cowper's gland ____________________
    e. hernia ____________________
    f. gynecology ____________________
    g. hermaphrodite ____________________
    h. cryptorchidism ____________________
    i. vasectomy ____________________
    j. castration ____________________

Label all of the features indicated on the photograph.

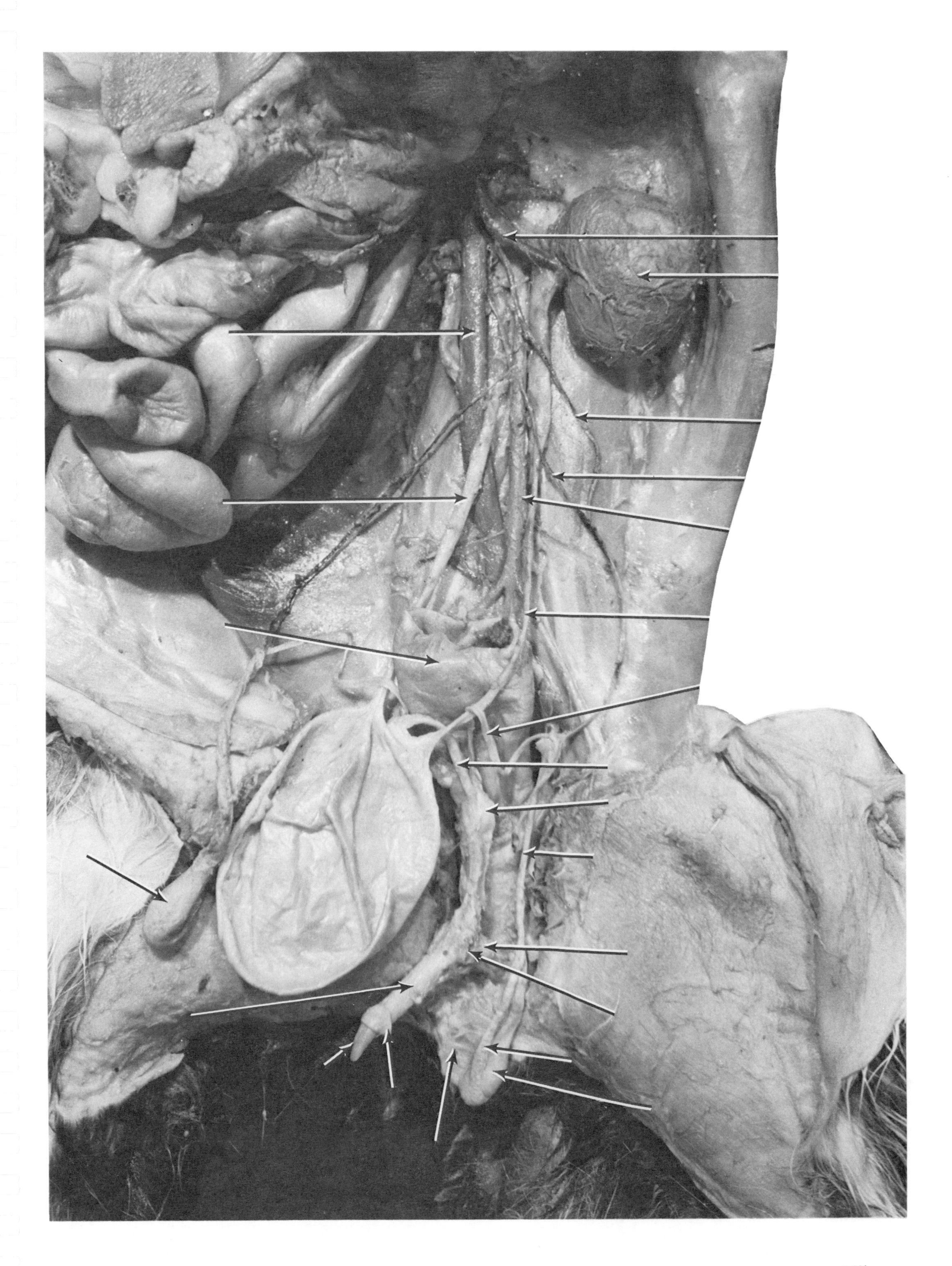

# THE NERVOUS SYSTEM: INTRODUCTION

We shall begin our study with an overview of the nervous system. It will be followed by the most delicate dissection you will be called upon to make, to find and expose narrow cranial and spinal nerves. The brain and spinal cord of the cat will be exposed. The large sheep brain will also be examined in detail. A study of two specialized sense organs, the eye and the ear will complete this unit.

**Overview**

The nervous system is concerned with communications by means of nerve impulses between the various parts of the body. It consists of the *central* and *peripheral* nervous systems.

The central nervous system refers to the *brain* and the *spinal cord,* while the peripheral nervous system refers to the nerve pathways and sense organs located outside the central nervous system.

The basic units of the nervous system are the nerve cells, or *neurons.* Their function is to receive stimuli, interpret sensations, and transmit messages from the central nervous system to other parts of the body.

**Types of Neurons** - There are three types of neurons.

1. **Sensory (Afferent) Neurons** - These receive stimuli and are found within *sense organs* or *receptors.* They transmit these sensations to the bain or spinal cord.
2. **Interneurons (Association Neurons)** - These neurons lie within the central nervous system and transmit messages up and down the spinal cord and between the different areas of the brain.
3. **Motor (Efferent) Neurons** - They carry messages from the central nervous system to the *effectors* (muscles or glands) that will respond to the stimulus.

**Parts of a Neuron** - The diagram below indicates the structure of a typical motor neuron.
It consists of the:

**Cell Body** - This is the enlarged portion of the neuron containing the nucleus. It is located within the gray matter of the spinal cord or brain.

**Dendrites** - These are thin protoplasmic extensions protruding from the cell body. They are highly branched and serve as reception areas for signals coming to the neuron from other cells.

**Axon** - This is the long narrow extension that carries impulses away from the cell body.

**Myelin Sheath** - This lipid covering protects the narrow axon. It is composed of cell membranes of *Schwann cells* repeatedly wrapped around the axon.

**Schwann Cells** - These nucleated cells form the outer boundary of the myelin sheaths, the *neurilemma.*

**Nodes of Ranvier** - These are gaps in the myelin sheath that occur at the junctions of adjacent Schwann cells.

**Axon Terminals** - These are non-myelinated branched endings of the axons.

**Synaptic Knobs** - These button-like terminals of the axon are usually embedded within muscular or glandular tissue.

The area where nerve tissue interacts with muscle tissue is known as the *neuro-muscular junction*. It is here that the nerve impulse from the central nervous system is translated into action. When the impulse reaches the synaptic knobs *transmitter substances* are released. These are chemicals such as *acetylcholine* and *epinephrine* which initiate changes resulting in contraction or secretion.

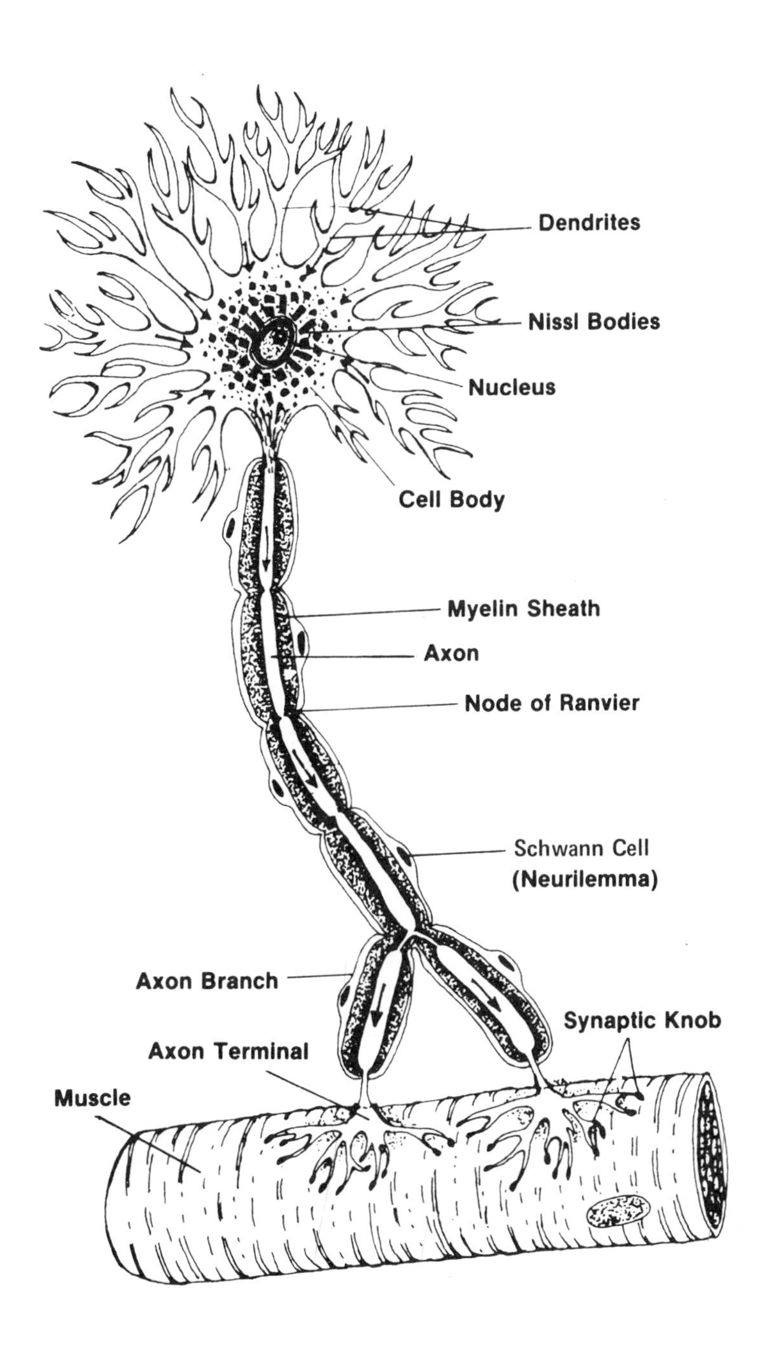


MOTOR NEURON

# NERVOUS SYSTEM
# CERVICAL AND BRACHIAL REGIONS

It is assumed that during the earlier dissections you have left intact the nerves associated with the structures studied.

Most nerves are extremely fine and delicate, and unlike arteries and veins nerves have not been stained. Thus their identification will be difficult. The best dissection instrument for clearing and tracing nerves is the dissecting needle. Most of your work will involve meticulously picking away muscle, fat, and connective tissue from around nerves.

There are twelve pairs of *cranial nerves* and thirty-eight pairs of *spinal nerves* in the cat. At this point we are interested in the last three of the cranial nerves only since these project into the cervical region.

Locate the anterior end of the *right common carotid artery* right below the jaw. Three cranial nerves are seen as in the photo. These are:

**Hypoglossal** (Cranial Nerve #12) - This nerve originates from the *medulla* of the brain. In gives branches to some ventral neck muscles and to the tongue.

**Spinal Accessory** (Cranial Nerve #11) - This nerve arises from the medulla and anterior spinal cord. It supplies the cleidomastoid, sternomastoid, and trapezius muscles.

**Vagus** (Cranial Nerve #10) - This is the longest of the cranial nerves. It too originates from the medulla. It sends branches to the ear, pharynx, larynx, lungs, heart, esophagus, stomach, and other abdominal viscera as far as the transverse colon.

We first observe the *vagus nerve* at the base of the jaw at the *nodose ganglion.* It lies slightly lateral of the *superior* cervical ganglion. The two sets of ganglia are, however, difficult to distinguish due to their close association.

The *vagus nerve* passes posteriorly, lying laterad to the *common carotid artery.* It is closely bound to the cervical portion of the *sympathetic nerve trunk.*

**Sympathetic Trunk** - The cervical sympathetic nerve trunk originates at the *superior cervical ganglion.* In the photo, the connective tissue between the vagus and the sympathetic trunk has been removed and they are clearly seen as separate nerves. The two accompany one another into the thorax.

**Spinal Nerves** - There are eight *cervical* and thirteen *thoracic spinal nerves.* After each spinal nerve leaves the vertebral column through the *intervertebral foramen* it divides to form two *rami,* dorsal and ventral. The *dorsal* ramus is smaller, supplying the dorsal body wall and skin. The larger *ventral ramus* supplies the ventral skeletal muscles, the arms, and legs. Many of the ventral rami unite in a complex manner to form networks, or *plexuses,* before being distributed to the musculature and skin. This is especially true in the region of the appendages.

**Brachial Plexus** - The brachial plexus is such a network, innervating the muscles of the upper limbs and thorax. It is formed by the ventral rami of the last four cervical nerves and the first thoracic nerve. The brachial plexus lies mediad to the shoulder and anterior to the first rib. Some of the nerves extending from the brachial plexus are the:

**Suprascapular** - This nerve is the most anterior of the brachial plexus group. It originates from the sixth cervical nerve and extends over the anterior edge of the scapula to supply the supraspinatus muscle.

**Subscapular** - Three subscapular nerves emerge from cervical nerves number six, seven, and eight to innervate the subscapularis, teres major, and latissimus dorsi muscles

**Phrenic** - This nerve leads to the diaphragm and is therefore most vital to respiration. In the cat it originates from slender branches of the fifth and sixth cervical nerves which unite to form the *phrenic nerve*. The entire nerve, including its cervical origins, is clearly visible in the photo on page 143.

**Axillary** - This nerve arises from cervical nerves number six and seven, extends laterally to supply the long head of the triceps brachii and the spinodeltoid muscles.

**Radial** - This is the largest nerve of the brachial plexus. It first supplies the triceps brachii, then continues down the forelimb as the *superficial radial nerve*.

**Median** - This nerve supplies most of the ventral muscles of the lower forelimb.

**Ulnar** - This is the most posterior nerve to pass down the forelimb. It lies close to the axillary and brachial arteries. It supplies the flexor capri ulnaris muscle, other muscles of the lower forelimb, as well as muscles in the paw.

**Ventral Thoracic** - These nerves arise from the eighth cervical and first thoracic nerves. They supply the underside of the pectoralis muscles. These nerves and many of their fine branches are clearly seen in the photo.

**Long Thoracic** - This nerve arises from the seventh cervical nerve, passes along the lateral thoracic wall upon the serratus ventralis muscle which it supplies.

---

**Label the cranial and spiral nerves of the human in the diagram below and compare them with those you observed in the cat.**

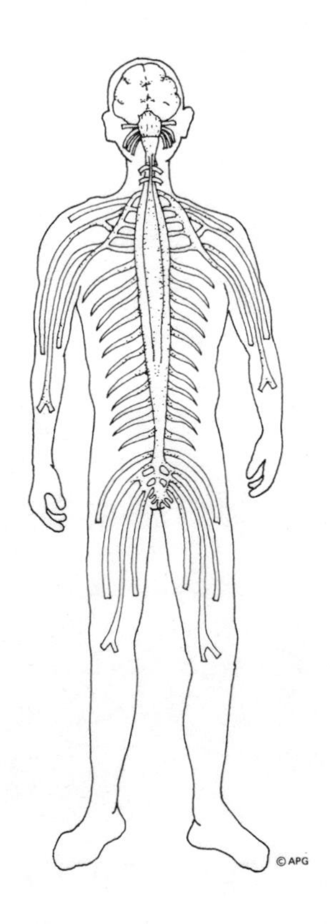

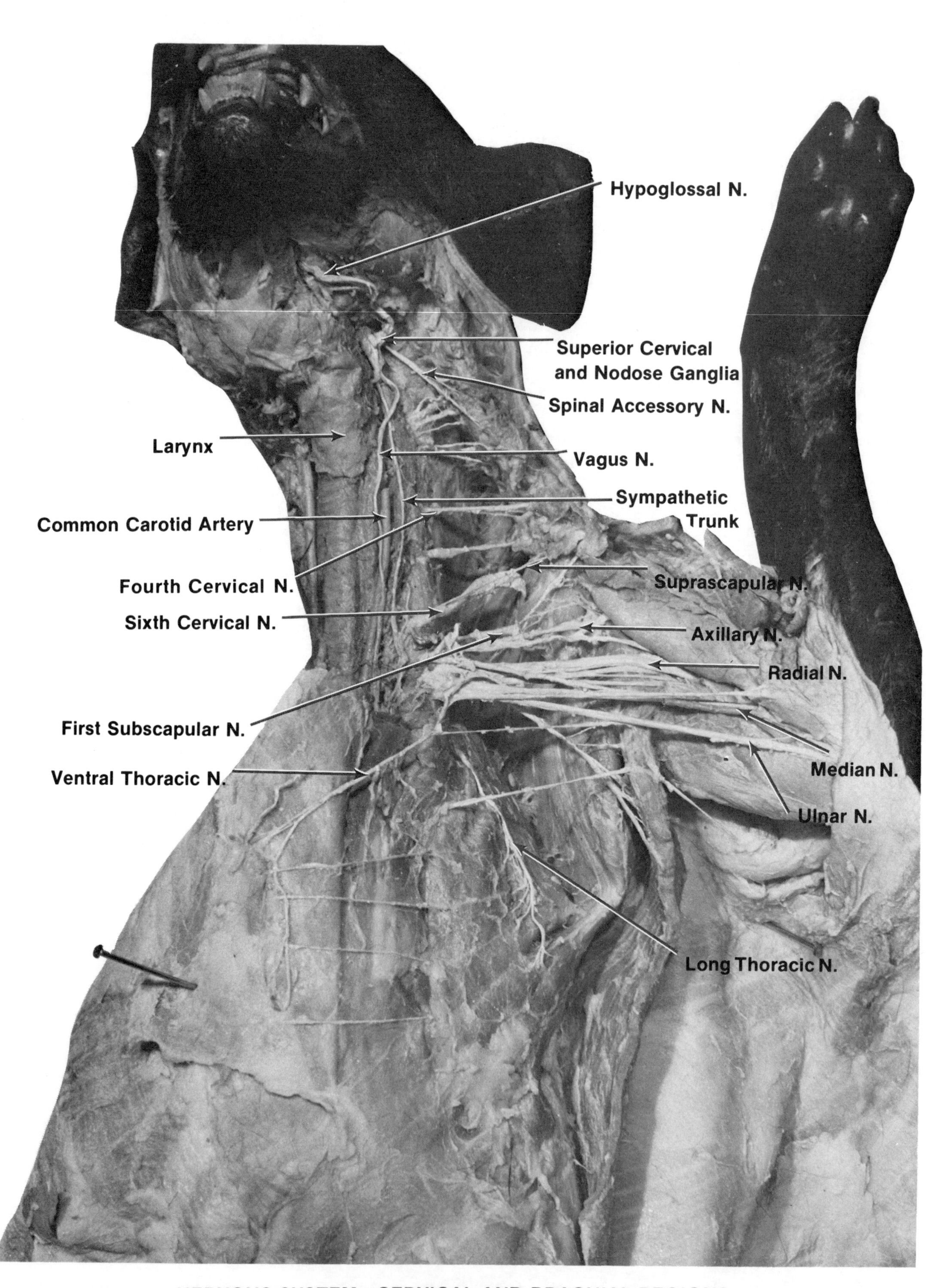


NERVOUS SYSTEM - CERVICAL AND BRACHIAL REGIONS

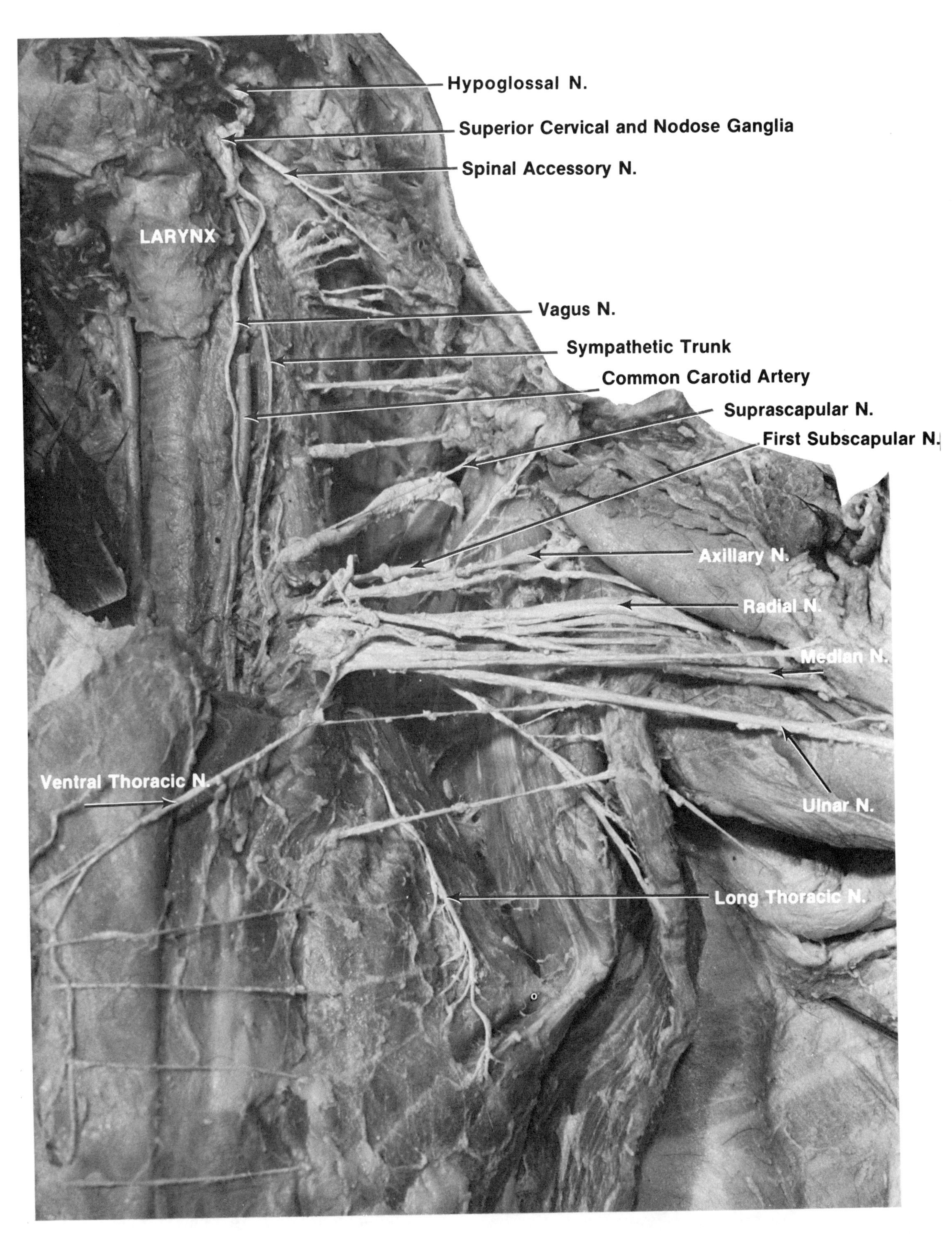


NERVOUS SYSTEM - CERVICAL AND BRACHIAL REGIONS (CLOSE-UP)

# NERVOUS SYSTEM - THORACIC REGION

If the thorax of your specimen has not yet been opened, do so now.

Use your scissors to cut through the ventral musculature, and the cartilaginous portion of the ribs. Begin by making an incision at the base of the rib cage about a half inch to the side of the xiphoid process. Continue to cut cranially along the side of the sternum through the top rib. Press the rib cage back dorsally to expose the entire thorax. Remove the left lung. Expose the nerves as seen in the accompanying photo.

Note that when the rib cage is bent dorsally as shown, some of the cervical nerves are still visible, but the brachial plexus is not.

**Phrenic** - Locate the *phrenic nerve*. In this photo it has been exposed from its origin in the fifth and sixth cranial nerve, through the thorax, closely associated with the pericardial membranes, to its tripod terminal on the anterior surface of the *diaphragm*. Most of the phrenic nerve lies within the thorax.

Return to the cervical area and locate the *spinal accessory nerve*, the *nodose* and *superior cervical ganglia*, and the *cervical sympathetic trunk*. All of these were studied in the last section.

**Vagus Nerve** (in thorax) - Continue tracing the vagus nerve caudally through the thorax. Within the thorax the right vagus lies along the trachea, while the left vagus lies along the esophagus. Near the subclavian artery the vagus gives off the *recurrent laryngeal nerve* which passes cranially to the larynx. Near the root of the lung the vagus gives off a network of branches to the heart and lungs. It continues caudally by dividing into the *dorsal and ventral branches* which lie along the esophagus. The right and left ventral branches join and near the diaphragm the right and left dorsal branches join. Thus, instead of a right and left vagus, a *ventral vagal trunk* and *dorsal vagal trunk* have formed. Both trunks of the vagus pass through the diaphragm along with the esophagus.

**Sympathetic Nerve Trunks** - These consist of a double line of ganglia and interconnecting nerves which lie on either side of the vertebral column. They extend from the base of the skull to the tail. Delicate *rami* connect the ganglia to the spinal nerves.

We have already studied the origin of the *sympathetic trunk* at the *superior cervical ganglion*. In the neck it was closely associated with the vagus nerve and common carotid artery. Within the thorax the sympathetic trunk separates from the vagus. It enters the middle cervical ganglion at the first rib, and more medially at the level of the second rib it joins the *inferior cervical (stellate) ganglion*. From here branches are given off to the heart. Before passing through the diaphragm the sympathetic trunk gives off the *greater and lesser splanchnic nerves*. These synapse with ganglia below the diaphragm.

**Thoracic Spinal Nerves** - Also seen in the photo are the nerves between the ribs, in the *intercostal* area. They are the *thoracic nerves*, thirteen pairs in all. The first thoracic nerve contributes to the brachial plexus. The ventral rami of the remaining thoracic nerves are called *intercostal nerves*. They follow closely the path of the intercostal arteries and veins. Remove some of the intercostal muscle between two ribs and observe the relationship between the intercostal artery, vein, and nerve.

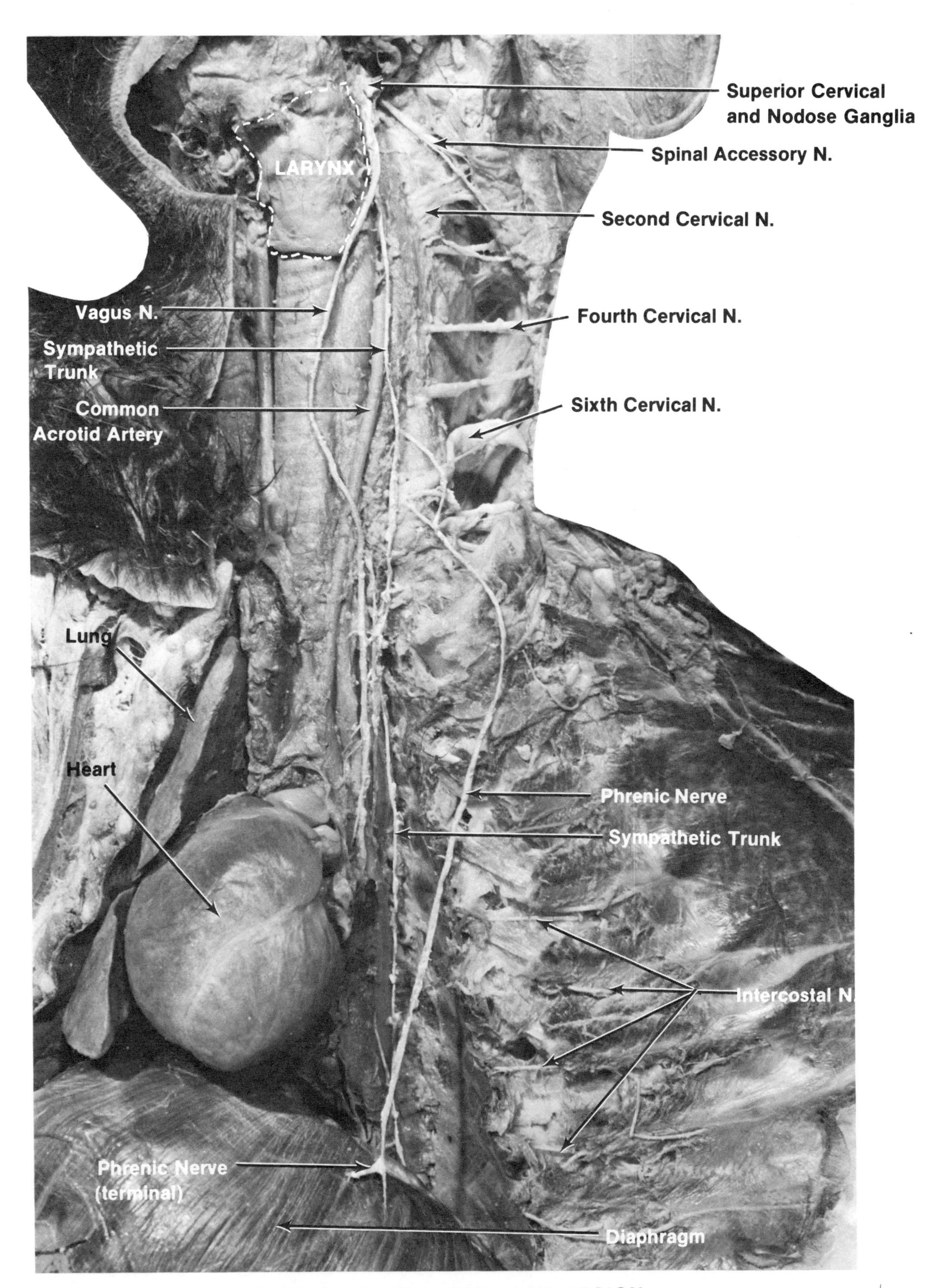


NERVOUS SYSTEM - THORACIC REGION

# NERVOUS SYSTEM - ABDOMINAL REGION

We shall first examine the nerves in the abdominal region. Again, the *vagus nerve* and the *sympathetic trunk* are the dominant nerves in this area. See the photo on page 147.

**Vagus Nerve** (in abdominal region) - It was already pointed out that the *ventral trunk* and *dorsal trunk* of the vagus nerve pierce the diaphragm along with the esophagus. The first branches supply the stomach. Vagus fibers, primarily from the dorsal trunk, pass to the *celiac plexus*, a network of nerves near the celiac artery, to supply the abdominal viscera as far posteriorly as the transverse colon. Viscera below this region receive parasympathetic innervation from sacral spinal nerves.

**Sympathetic Trunk** - (in abdominal region) - We have noted that anterior to the diaphragm the sympathetic trunk gives off the *greater splanchnic* and *lesser splanchnic nerves*. These contain preganglionic fibers that synapse with the *celiac ganglion* and the *superior mesenteric ganglion*. These ganglia lie near the base of the arteries of the same name. Together they constitute the *solar plexus*. From here many fine nerve fibers branch to the abdominal viscera.

The two parallel sympathetic trunks continue caudally with small ganglia along the chain. Some preganglionic fibers synapse with the *inferior mesenteric ganglion*, at the base of the inferior mesenteric artery, while postganglionic fibers follow the artery and its branches to the viscera. They terminate in the sacral region.

# NERVOUS SYSTEM - THE LUMBOSACRAL PLEXUS

In the photo on page 147, the kidney, horn of the uterus, and the urinary bladder have been moved to the right. Parts of the dorsal musculature, the iliopsoas and psoas minor, have been removed to expose the nerves in the lumbar and sacral areas.

**Lumbar and Sacral Spinal Nerves** - There are seven pairs of *lumbar* and three pairs of *sacral spinal nerves.* The first three lumbar nerves divide into *lateral and medial branches* to supply the muscles of the lateral and ventral muscle walls. The ventral rami of the last four lumbar nerves and the three sacral nerves are interconnected to form a network known as the *lumbosacral plexus.*

Some of the nerves originating at the *lumbosacral plexus* are the:

**Genitofemoral** - This nerve originates from the fourth and fifth lumbar nerve. It supplies the integument of the external genitalia, the medial side of the thigh, and the adjacent body wall.

**Lateral Cutaneous** - This nerve also originates from the fourth and fifth lumbar nerves. It supplies the skin on the lateral surface of the thigh.

**Femoral** - This large nerve is easily seen. It is formed by the fifth and sixth lumbar nerves. It emerges from the psoas major muscle and distributes to the ventral femoral muscles.

**Obturator** - This nerve originates from the sixth and seventh lumbar nerves. It passes posteriorly through the obturator foramen to supply the medial femoral muscles, the adductors, pectineus, and gracilis.

**Lumbosacral Cord** - This tough nerve band from the sixth and seventh lumbar nerves passes posteriorly. It joins the last two lumbar nerves to the sacral nerves.

In order to expose the lumbosacral cord as well as the remainder of the nerves of the lumbosacral plexus turn the cat over with its dorsal surface upward and dissect the lateral thigh muscles.

Cut through the middle of the *biceps femoris* and reflect both halves. Remove the tenuissimus and tensor fascia lata muscles. It is necessary to also cut the caudofemoralis, gluteus maximus and gluteus minimus muscles in order to expose the nerves in this area. Your dissection should now appear as in the photo on page 148.

**Sciatic Nerve** - This very large nerve was already observed during the dissection of the muscles of the thigh. It is about ¼ inch thick and passes below the muscles of the lateral surface of the thigh and gluteal area. The *sciatic nerve* is formed from the *lumbosacral cord* and the *first sacral nerve.* It also receives branches from the second sacral nerve. It gives off a number of nerves to the thigh and lower leg muscles. They include the:

> **Muscular Branch** - Near its point of exit from the pelvis a large muscular branch divides from the sciatic nerve. It supplies the biceps femoris, semitendinosus, semimembranosus, and adductor femoris muscles.

**Sural** - This nerve branch of the sciatic nerve is given off above the knee. It passes to the gastrocnemius muscle, extending as far as the ankle

**Common Peroneal and Tibial** - Near the knee the sciatic nerve divides to form the *common peroneal nerve* and the *tibial nerve*. These major nerves lead to the muscles of the lower leg.

Other nerves in the sacral region include the:

**Gluteal** - The *superior gluteal nerve* arises from the lumbosacral part of the spinal cord, passes laterally to supply the gluteus medius and minimus muscles. The *inferior gluteal nerve* arises from the first sacral nerve and the lumbosacral cord. It passes caudally to the femur to supply the caudofemoralis and gluteus maximus muscles.

**Pudendal** - This nerve arises from the second and third sacral nerves and passes caudally to supply the anus and external genitalia.

**Posterior Femoral Cutaneous** - This nerve accompanies the inferior gluteal blood vessels and passes into the thigh. It arises from the second and third sacral nerves to supply the anal region and the biceps femoris muscle. It extends along the thigh as far as the popliteal space.

**Inferior Hemorrhoidal** - This nerve originates from the second and third sacral nerves. It passes to the urethra and urinary bladder and to the caudal end of the rectum.

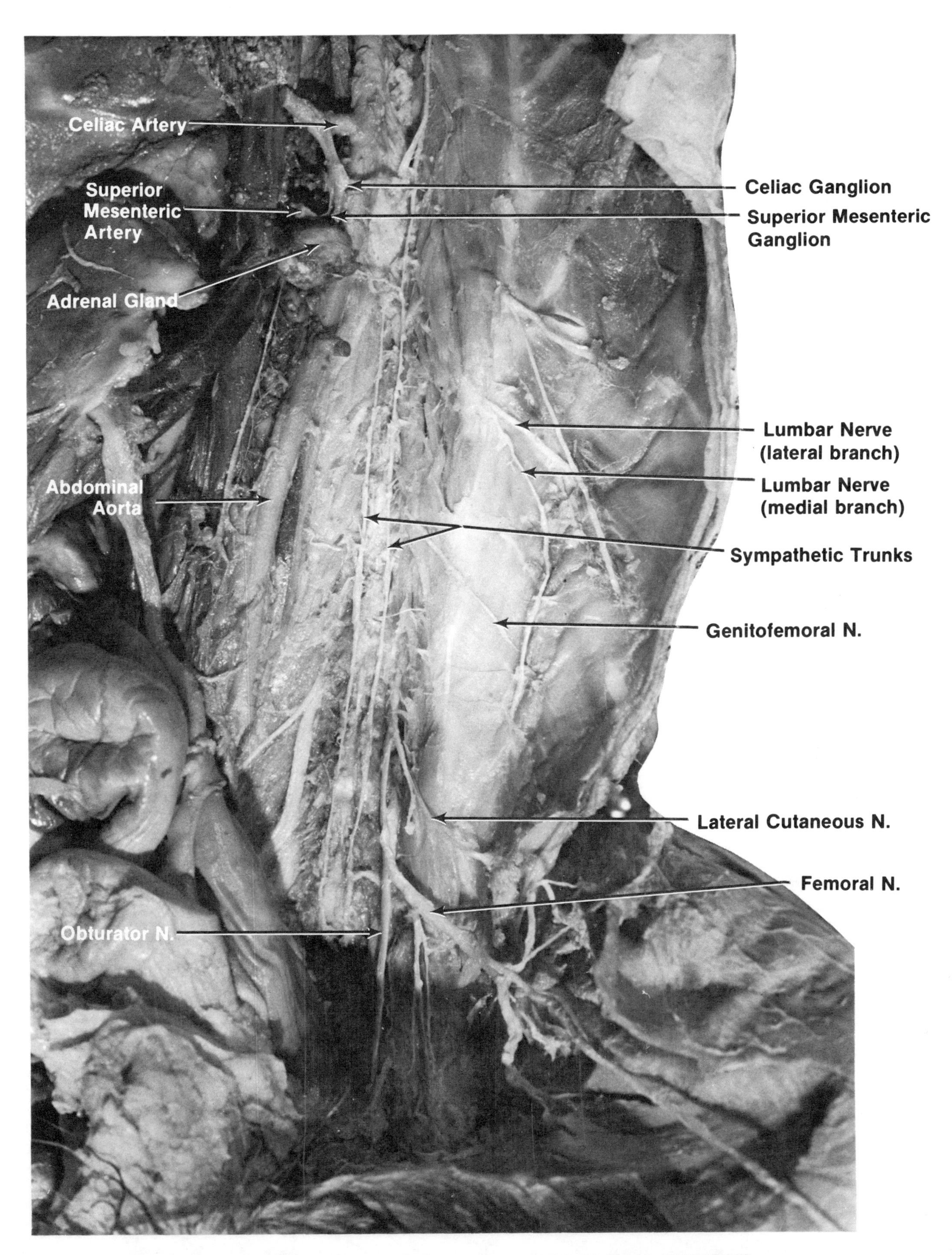


NERVOUS SYSTEM - ABDOMINAL AND LUMBOSACRAL REGIONS

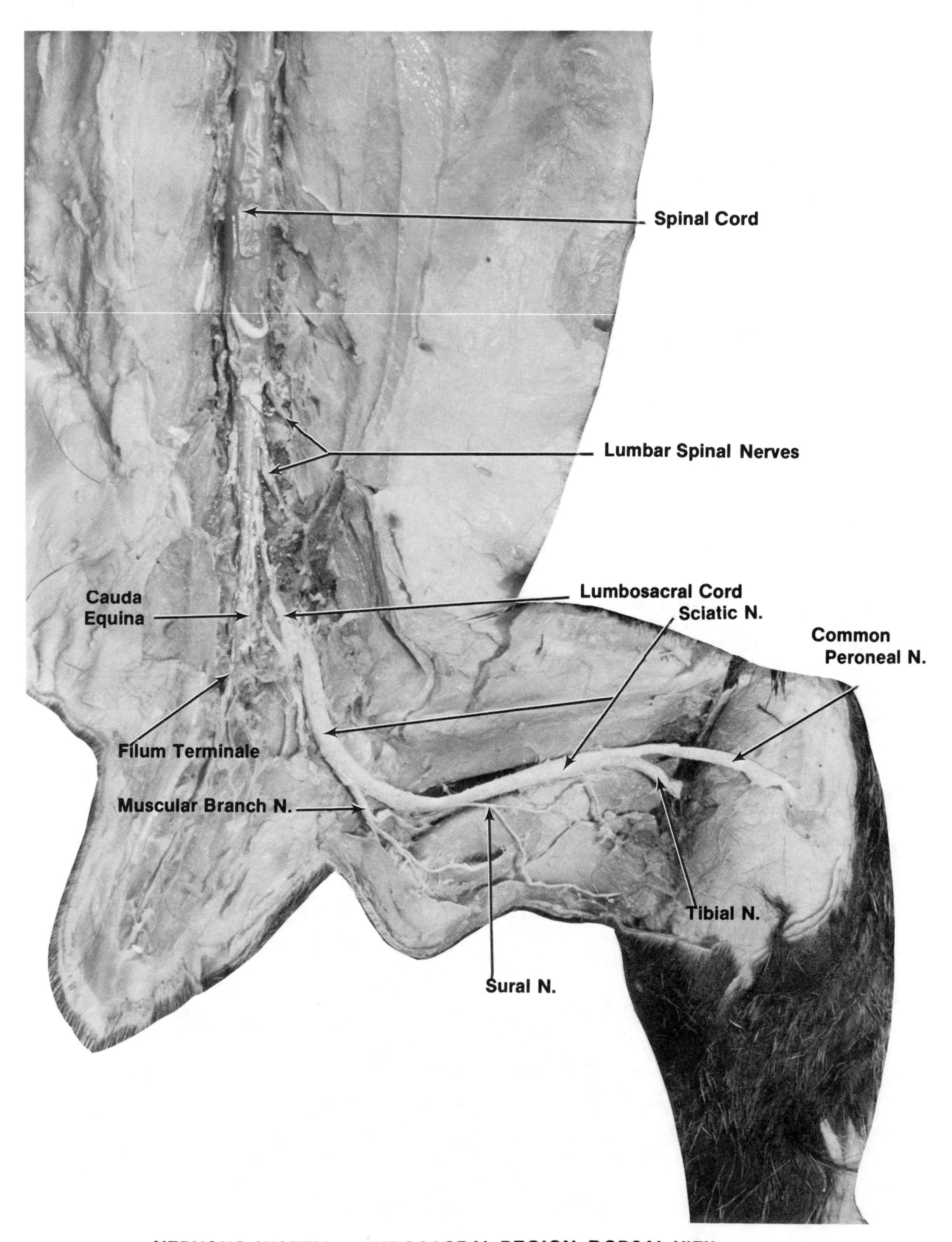


NERVOUS SYSTEM - LUMBOSACRAL REGION, DORSAL VIEW

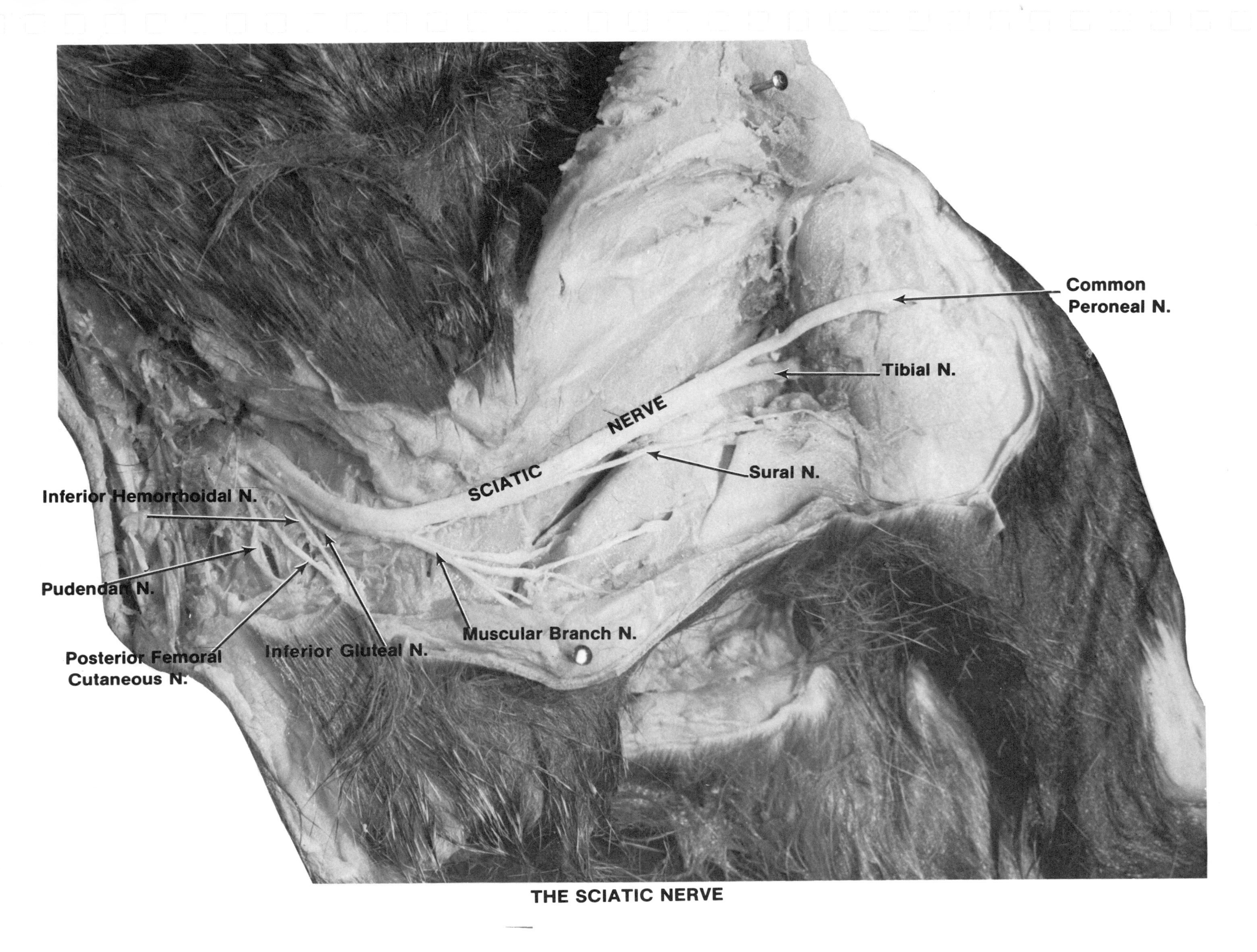


THE SCIATIC NERVE

Name ................................. Section ....... Date.................

## SELF - QUIZ VIII
## NERVOUS SYSTEM - CERVICAL TO CAUDAL NERVES

1. How many spinal nerves are there in the cat?
2. Name the segments of the spinal cord and the number of spinal nerves in each segment.
3. Which spinal nerves comprise the brachial plexus and the lumbosacral plexus?
4. Name the longest cranial nerve. What are some of its terminals?
5. Which nerve leads to the diaphragm? Where does it originate?
6. Name the large nerve supplying the lateral aspect of the thigh.
7. Name the major nerves originating at the brachial plexus.
8. Name the major nerves originating at the lubosacral plexus.
9. Besides the cervical spinal nerves, what major nerves pass through the cervical area?
10. Define the terms listed below.

ANSWERS

1. ______________________________
2. ______________________________
3. ______________________________
4. ______________________________
5. ______________________________
6. ______________________________
7. ______________________________
8. ______________________________
9. ______________________________
10. a. stellate ganglion ______________________________
    b. solar plexus ______________________________
    c. sympathetic nervous system ______________________________
    d. parasympathetic nervous system ______________________________
    e. greater splanchnic nerve ______________________________
    f. cauda equina ______________________________
    g. ventral rami ______________________________
    h. dorsal root ganglion ______________________________
    i. sympathetic ganglion ______________________________
    j. superior cervical ganglion ______________________________

Label all of the features indicated on the photograph.

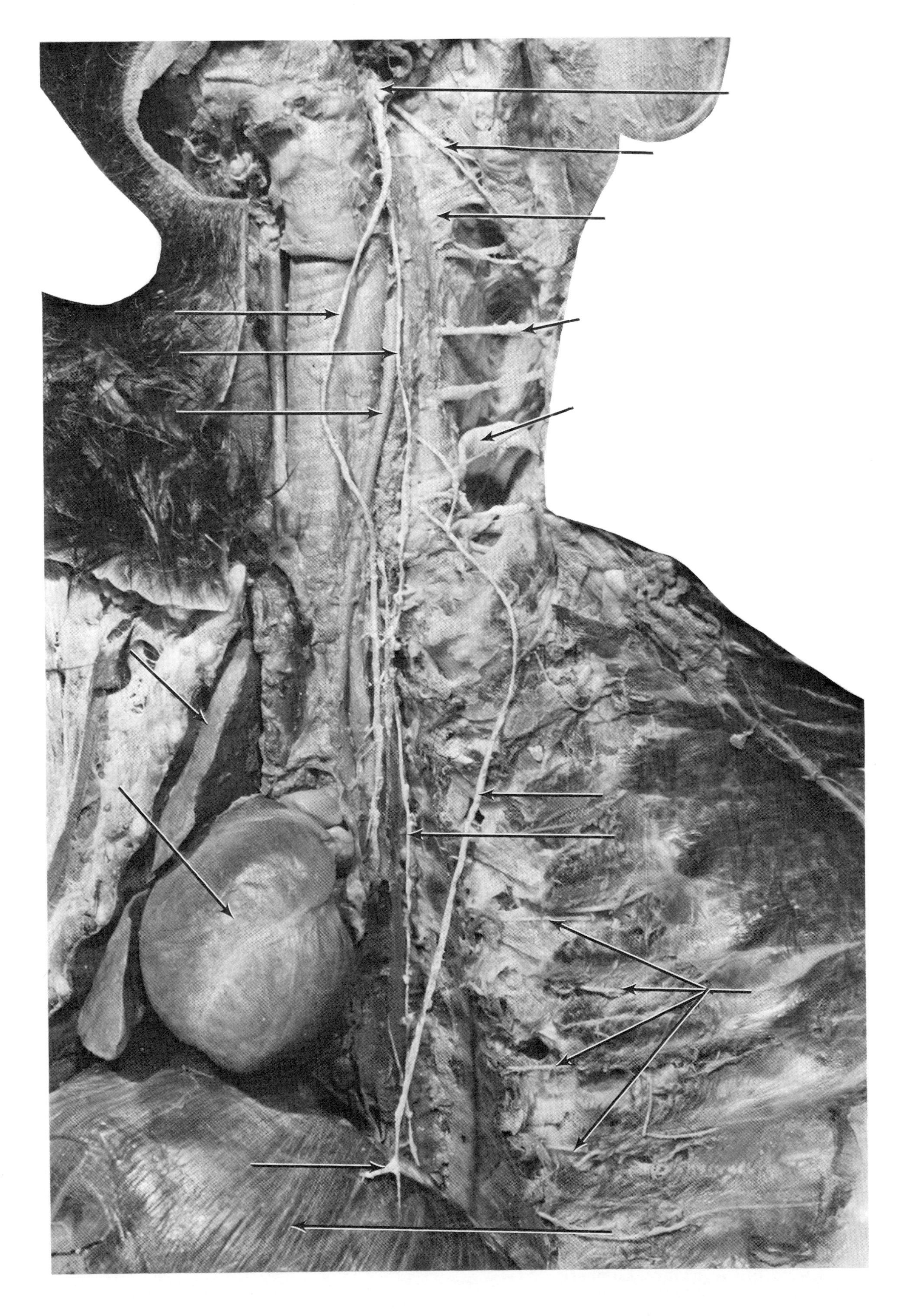

# NERVOUS SYSTEM - THE SPINAL CORD

The cat is placed on the dissection tray with the dorsal surface upward.

You have already begun to expose the *lumbar* and *sacral* nerves in the last section. The Photo on page 148 reveals the base of the spinal cord and the sciatic nerve. Your task is to expose the entire spinal cord, as in the next Photo, from the tail to the head.

Thirty-eight to thirty-nine pairs of *spinal nerves* arise from the spinal cord: eight *cervical,* thirteen *thoracic,* seven *lumbar,* three *sacral,* and seven to eight *caudal* nerves. In man there are only thirty-one pairs.

**Number of Spinal Nerves**

| | Cat | Man |
|---|---|---|
| Cervical | 8 | 8 |
| Thoracic | 13 | 12 |
| Lumbar | 7 | 5 |
| Sacral | 3 | 5 |
| Caudal (Coccygeal) | 7-8 | 1 |
| Total | 38-39 | 31 |

Begin to expose the spinal cord by first removing the muscles dorsal to the vertebral column. Then, using bone clippers, cut away the *neural arches* of the vertebrae to expose the spinal cord for its entire length. Work carefully, one vertebra at a time. Remove the neural arches from the sides of the vertebrae to expose the origins of the *spinal nerves.*

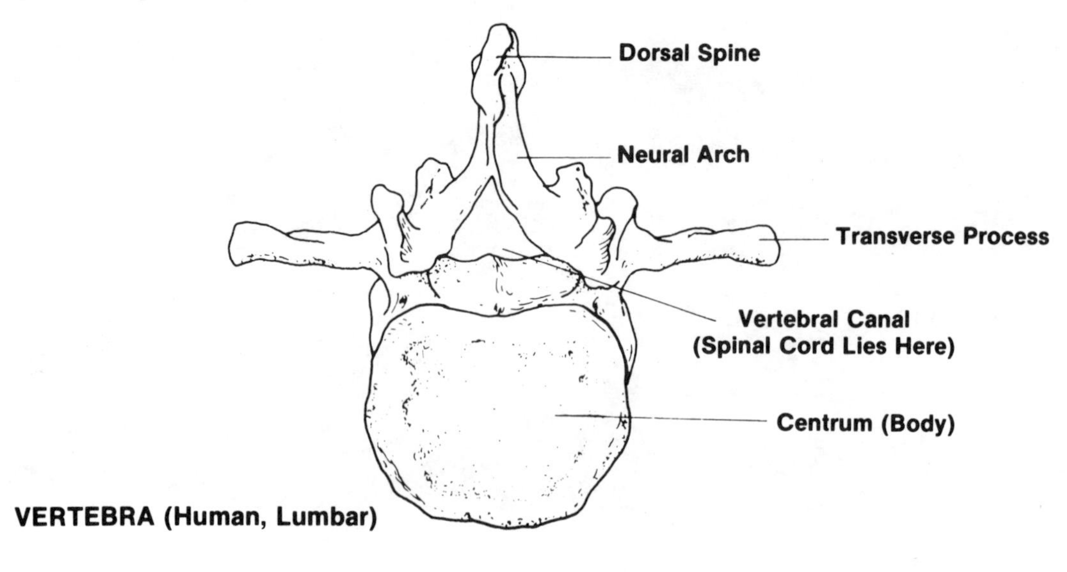


**VERTEBRA (Human, Lumbar)**

Each spinal nerve arises from two roots, the *dorsal root*, which is *sensory*, and the *ventral root*, which is *motor*. These unite a short distance from the cord to carry sensory and motor impulses to and from the spinal cord. For this reason, all spinal nerves are known as *mixed* nerves. Find the prominent rounded swellings on the dorsal root proximal to its union with the ventral root. These enlargements contain the cell bodies of the sensory neurons and are known as *dorsal root ganglia*.

Remove a ¼ inch section of the spinal cord and observe in cross section with a low power dissecting microscope. Locate the *gray matter* in the shape of a capital "H" near the center. It contains the cell bodies of the motor neurons. The white matter around the periphery of the spinal cord is composed of neurons carrying messages up and down the spinal cord, from the brain to muscles and glands. Specialized tracts of white matter communicate with different brain centers.

The *meninges* are membranous coverings for the spinal cord; the *dura mater, arachnoid,* and *pia mater.* They are similar to those described covering the brain.

Note the diameter of the cord along its length. It is thickest in the area of the limbs. The *cervical enlargement* is found near the anterior limbs, while the *lumbo-sacral enlargement* is found near the lower limbs. This corresponds to the many nerves controlling the limbs which originate here.

While the cervical and thoracic spinal.nerves exit the spinal column horizontally, those in the lumbar and sacral region pass posteriorly and leave the bony spinal column at a lower region. This causes the appearance of a multi-fibered, tail-like structure within the spinal cord. It makes up the *cauda equina*, literally, horse's tail. The last nerve filament, a remainder of the spinal cord, is called *filum terminal.*

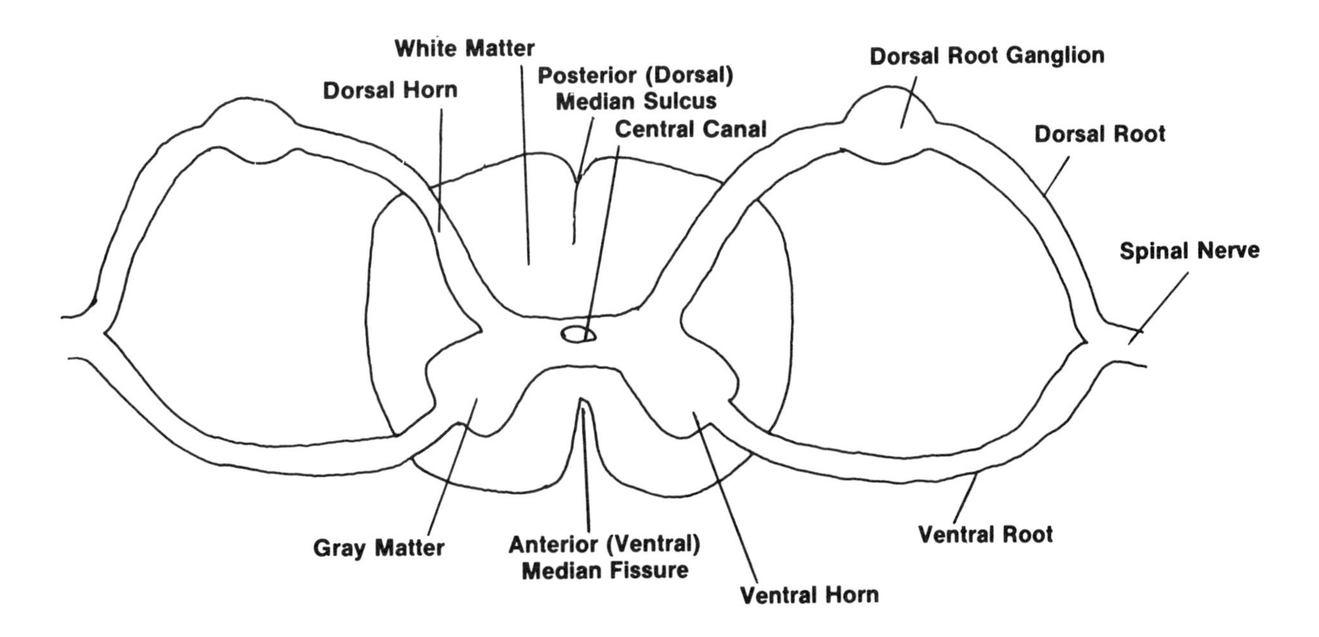


**HUMAN SPINAL CORD (Cross Section)**

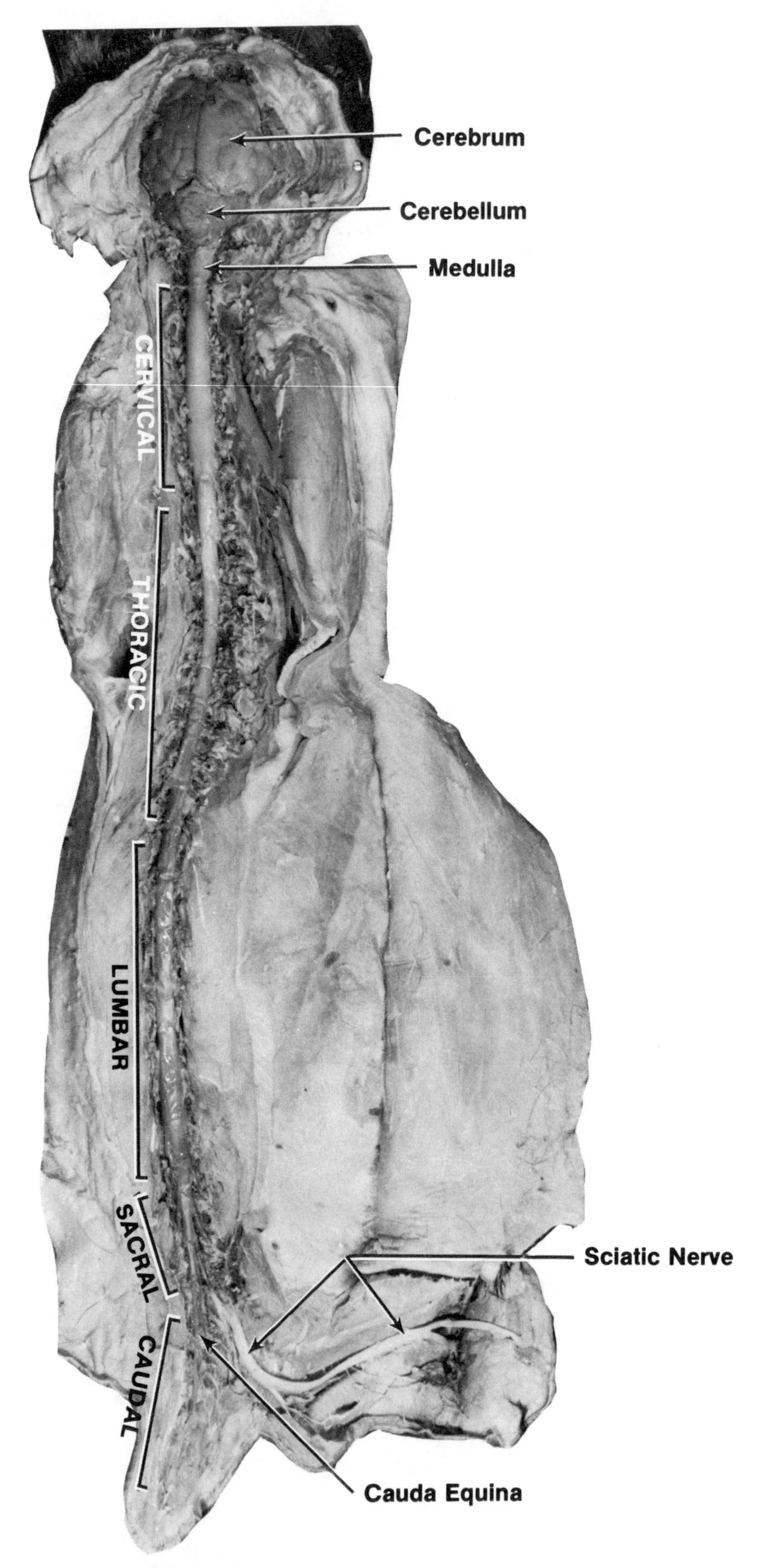


THE BRAIN AND SPINAL CORD

# NERVOUS SYSTEM - THE BRAIN

Remove the skin and the fibrous *epicranial aponeurosis* covering the top of the cat's skull. Expose the bony cranium.

Use a bone saw or bone clippers to make one mid-sagittal and several transverse cuts through the cranium. Use heavy forceps and insert between the bone and the membranous covering of the brain. With the forceps break the loosened chips of bone. Use extreme caution. Do not damage the delicate brain tissue.

Expose the cerebrum as far anteriorly as the heavy bony ridges above the nose and laterally to the level of the eyes. Remove more cranial bones posteriorly to expose the *cerebellum* and *medulla* as well as the top of the spinal cord. A transverse bony septum, the *tentorium* lies between the cerebrum and cerebellum. Be very careful in removing it. Do not injure or destroy the delicate brain tissue.

The brain is enclosed by three membranes collectively called the *meninges.* These are similar to and continuous with those of the spinal cord already studied. The outer, tough fibrous layer adheres closely to the inner surface of the bones of the cranium. It is called the *dura mater.* The delicate middle layer is the *arachnoid.* The inner, thin membrane, the *pia mater,* is highly vascular. It follows the contours and convolutions of the brain and adheres closely to it. Carefully remove the dura mater. During this procedure the arachnoid is also generally removed.

We shall now turn to a detailed study of the sheep brain. Carefully remove the brain of the cat in order to compare it to the much larger sheep brain.

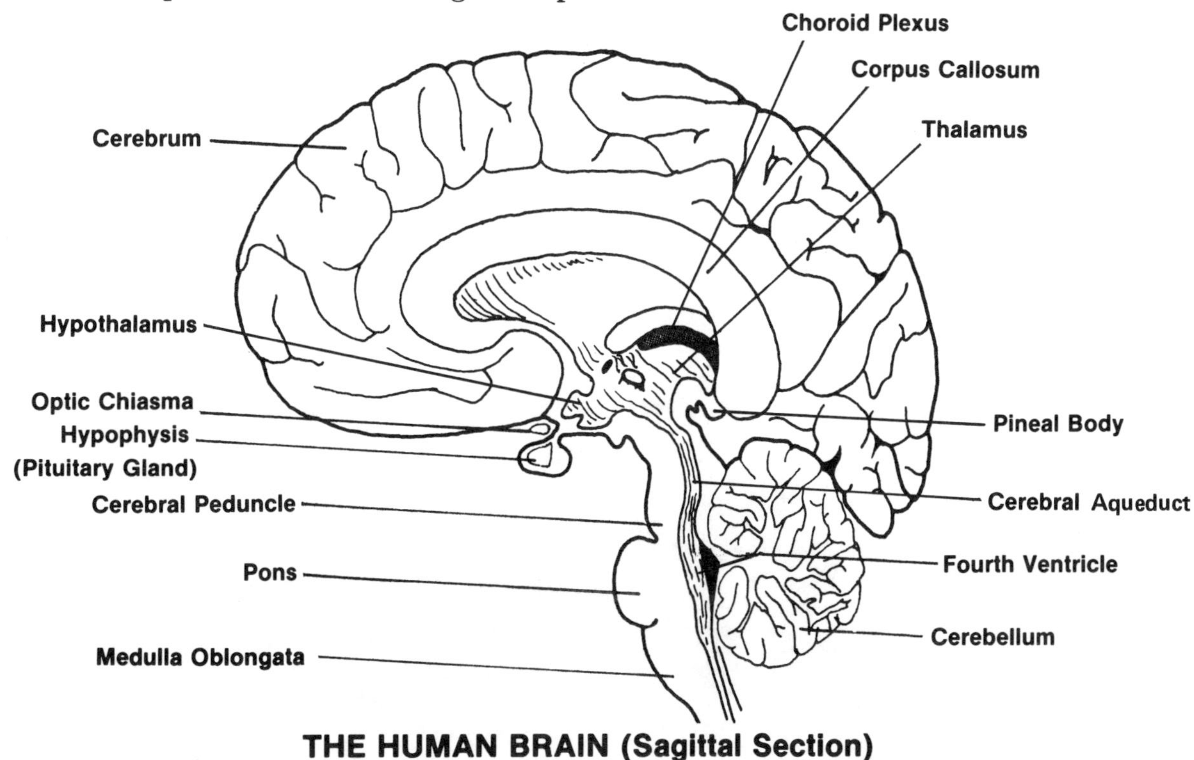


**THE HUMAN BRAIN (Sagittal Section)**

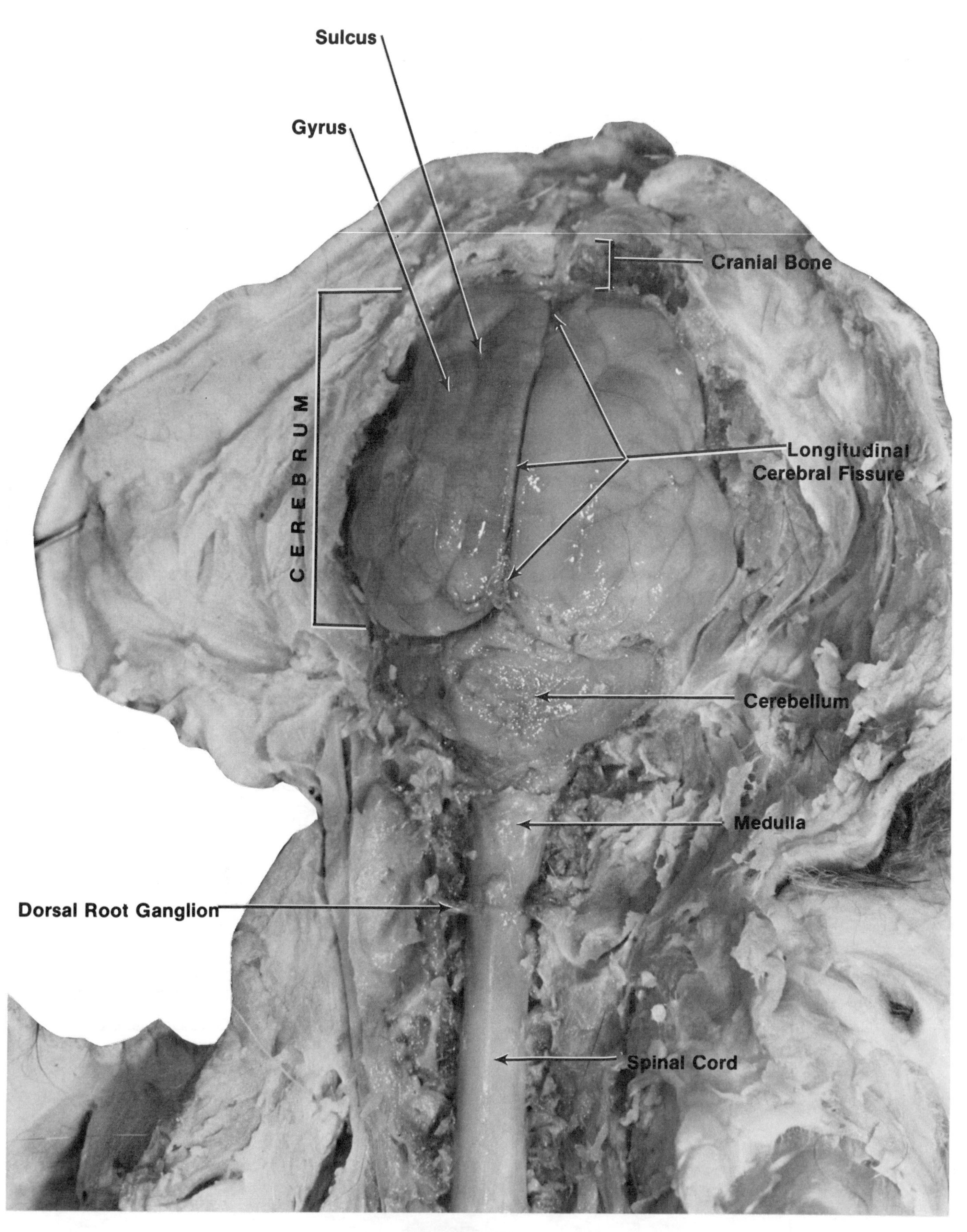
Sulcus
Gyrus
Cranial Bone
CEREBRUM
Longitudinal
Cerebral Fissure
Cerebellum
Medulla
Dorsal Root Ganglion
Spinal Cord

THE BRAIN

# SHEEP BRAIN

The sheep brain is very similar to that of the cat and human. Its large size makes it particularly well suited for study. Structures too small to be noticed in the cat brain may be readily identified in the sheet brain. Remove the exposed brain of the cat and compare it to that of the sheet throughout this exercise.

**Dorsal View** - Remove any portion of the *dura mater* still adhering to the surface of the brain.

Note the large *cerebral hemispheres* and their *convoluted* surfaces. The *sulci* are depressions and the *gyri* are raised areas of the surface. The right and left cerebral hemispheres meet at the *longitudinal cerebral fissure.*

You may easily identify the three major regions of the brain, the *cerebrum, cerebellum* and *medulla oblongata.*

**Ventral View** - More detailed structures are seen upon the ventral surface (underside) of the brain.

The sensory tract from the eye, the *optic nerve,* is seen in the photo, p. 160. Most of the fibers of this nerve cross to opposite sides of the brain thereby forming the "X" shaped *optic chiasma* clearly visible in the photo.

Terminals of other cranial nerves are also visible.

Note:

— The *olfactory bulb* at the anterior end of the cerebrum.

— The thickest of all of the cranial nerves, the *Trigeminal nerve,* originates at the anterior lateral surface of the *pons.*

— The *vagus nerve* is the longest of the cranial nerves. It exists from the lateral surface of the *medulla oblongata* and sends branches to the pharynx, larynx, heart, lungs, stomach and the intestines.

The brain is well supplied with blood. In the photo, p. 160, we see some of the arteries that join at the base of the brain (anastomosis) to form the *Circle of Willis.* Learn the names of these arteries.

The pituitary gland, or *hypophysis,* has been removed in this specimen. This often happens when the brain is forcibly removed from the cranium.

**Sagittal View** - Make a mid-sagittal section of the sheep brain. This view shows more details than the two seen previously. See photo, p. 161.

Note:

— The *Arbor Vitae* (Tree of Life) — This is the branched white matter within the *cerebellum.*

— The *Right Lateral Ventricle* — It can be seen bounded by the *corpus callosum* above and the *fornix* below. The corpus callosum is located at the base of the longitudinal cerebral fissure and connects the right and left cerebral hemispheres.

— The *Pineal Body.*

— The *Cerebral Aqueduct* — This passageway joined the third to the fourth ventricle.

— The *Fourth Ventricle.*

— The *Thalamus* (intermediate mass).

— The *Corpus Callosum* — This is a band of white fibrous tissue connecting the right and left halves of the brain. It forms a roof over the large *lateral ventricles* of the brain in which *cerebrospinal fluid* is found.

Locate and identify all of these structures in your specimen.

**Coronal Section -** Cut the brain in coronal section, perpendicular to the *longitudinal cerebral fissure.* Observe the difference in coloration between the outer ⅛ inch of brain tissue and that beneath. The darker tissue, or *gray matter,* is the *cortex.* Note that as a result of the extensive convoluted pattern of the cerebrum, the "gray" cortex extends deeply into the surface. The cortex contains the *cell bodies* of the brain's *neurons* while the white matter is composed of *myelinated axons.* Two thirds of the nerve cells of the entire body are located in the thin outer layer, the cortex, of the brain. Note too, the reversal of positions of gray and white matter between the brain and spinal cord. While the gray matter is toward the outer surface of the brain, it is centrally located in the spinal cord, surrounded by white matter (see p. 153).

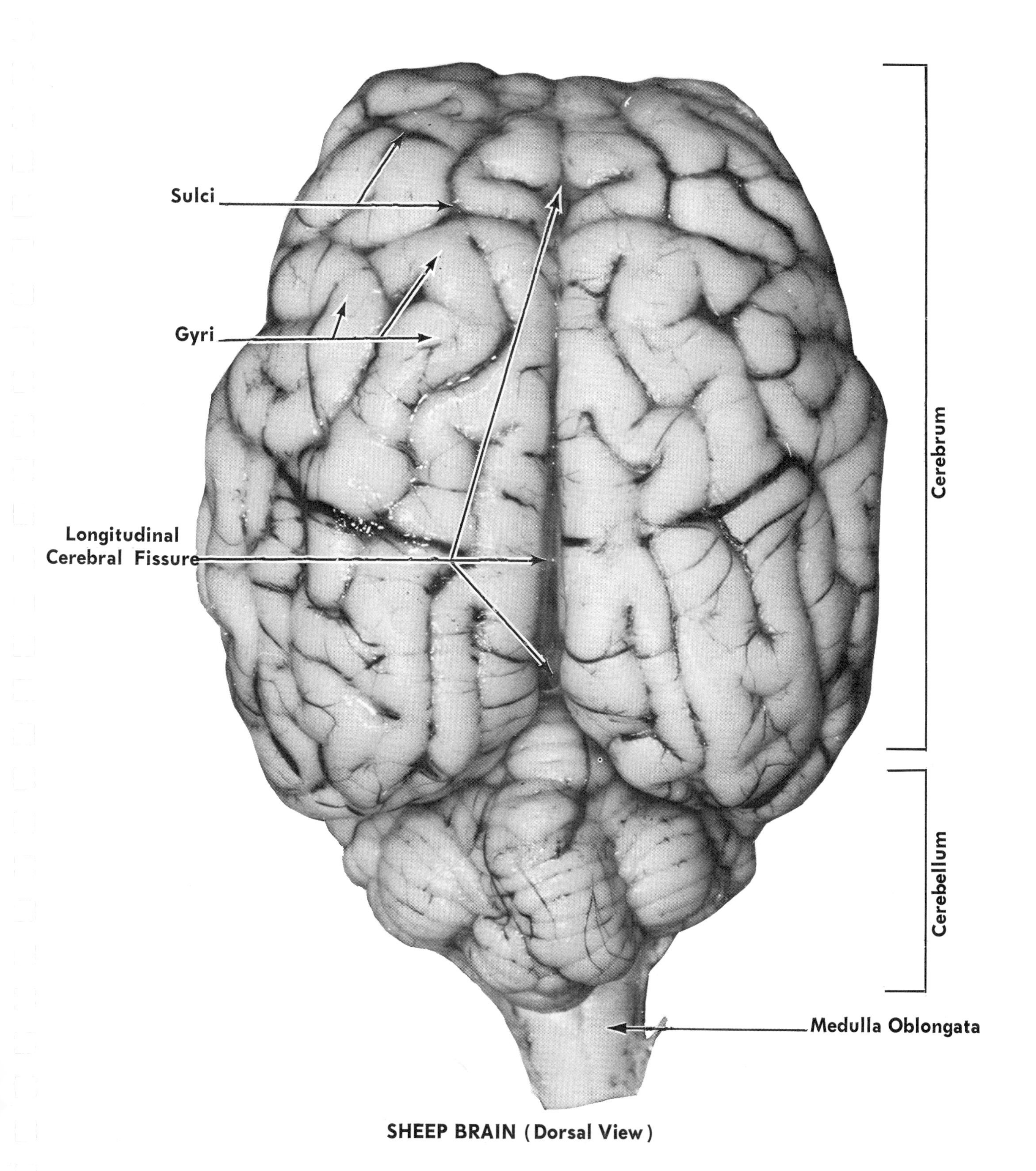


SHEEP BRAIN (Dorsal View)

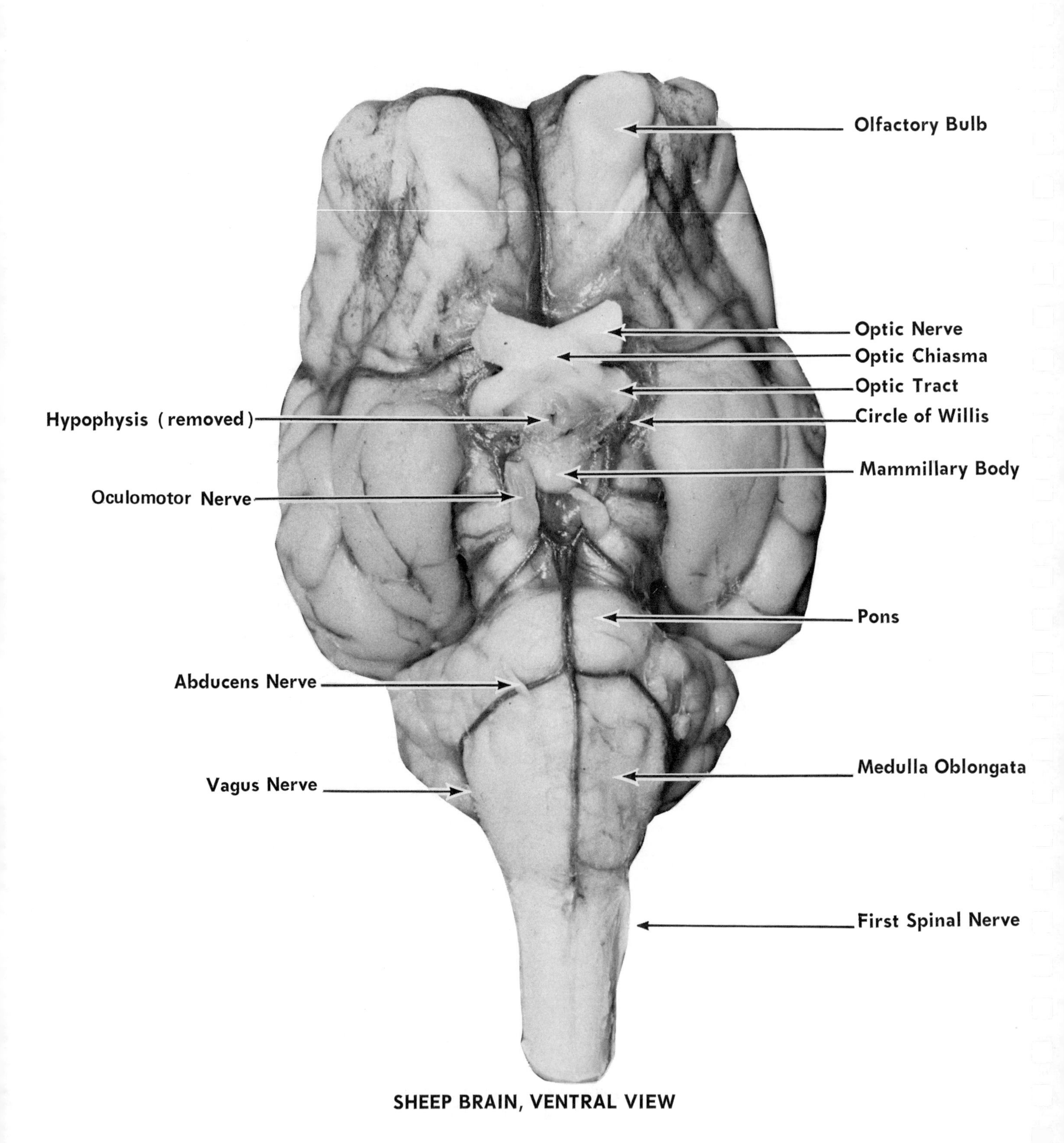


SHEEP BRAIN, VENTRAL VIEW

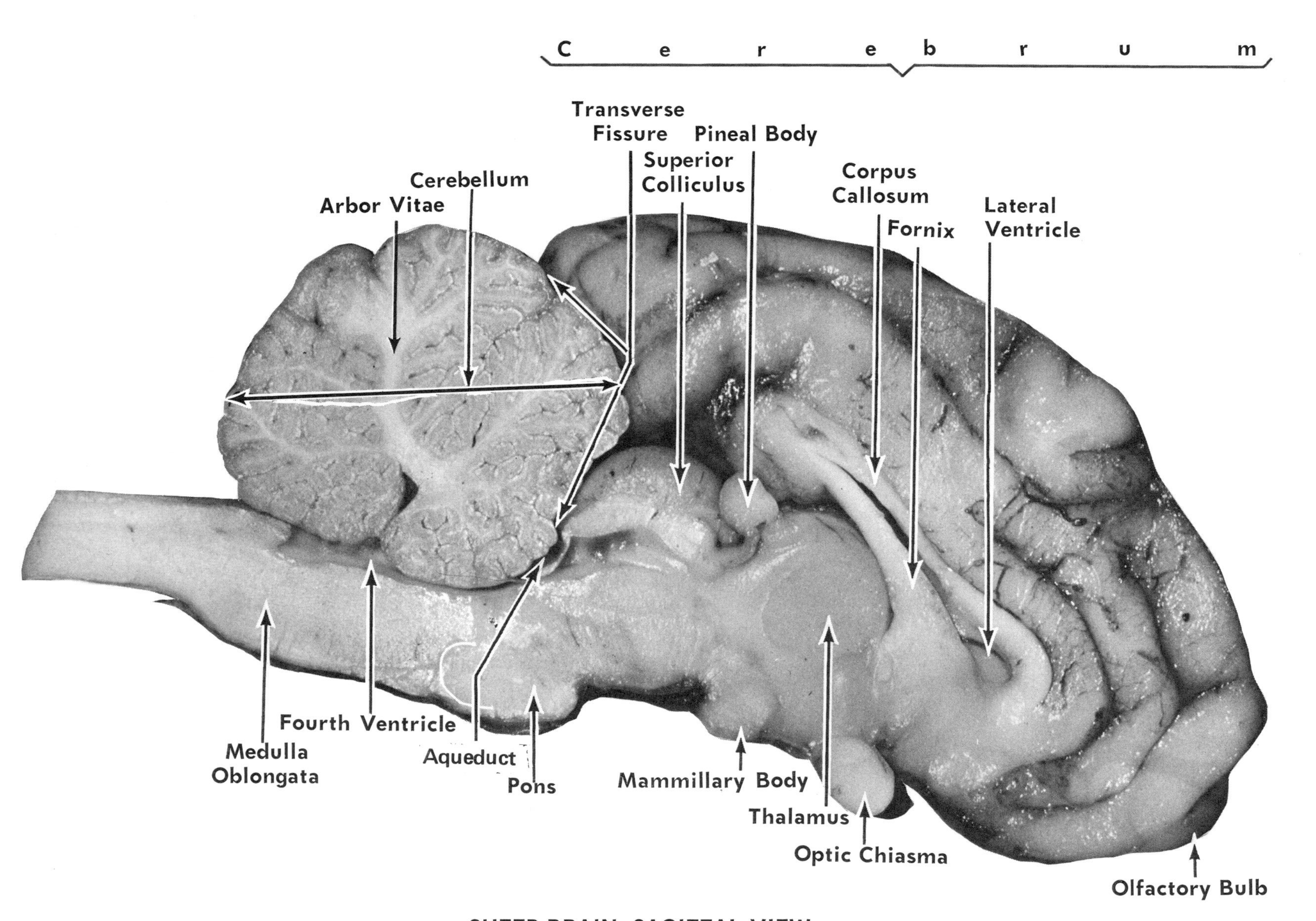


SHEEP BRAIN, SAGITTAL VIEW

Name .................................. Section ....... Date ................

# SELF - QUIZ IX
# BRAIN AND SPINAL CORD

1. The three major divisions of the brain are the a)________, b)________, and the c)_______ .
2. The folds upon the surface of the brain are known as ________________________________ .
3. The two hemispheres of the cerebrum are separated by the ____________________fissure.
4. The optic nerves from the eyes cross before entering the brain. The crossed structure is known as the ____________________.
5. From what part of the brain does the vagus nerve originate?
6. The hypophysis is more commonly known as the ______________________________ .
7. a. How many ventricles are there in the brain?
   b. What is contained within the ventricles of the brain?
8. The brain is protected by a portion of the skull known as the ______________________ .
9. The swellings on the dorsal roots of spinal nerves are known as ____________________ .
10. Define each of the terms listed below.

ANSWERS

1. (a) ____________________________________________
   (b) ____________________________________________
   (c) ____________________________________________
2. ____________________________________________
3. ____________________________________________
4. ____________________________________________
5. ____________________________________________
6. ____________________________________________
7. a. ____________________________________________
   b. ____________________________________________
8. ____________________________________________
9. ____________________________________________
10. a. meninges ____________________________________
    b. olfactory bulb ______________________________
    c. cranial nerve _______________________________
    d. corpus callosum _____________________________
    e. fornix ______________________________________
    f. cortex ______________________________________
    g. sulci _______________________________________
    h. gyri ________________________________________
    i. ganglion ____________________________________
    j. Circle of Willis ____________________________

Label all of the features of the photograph.

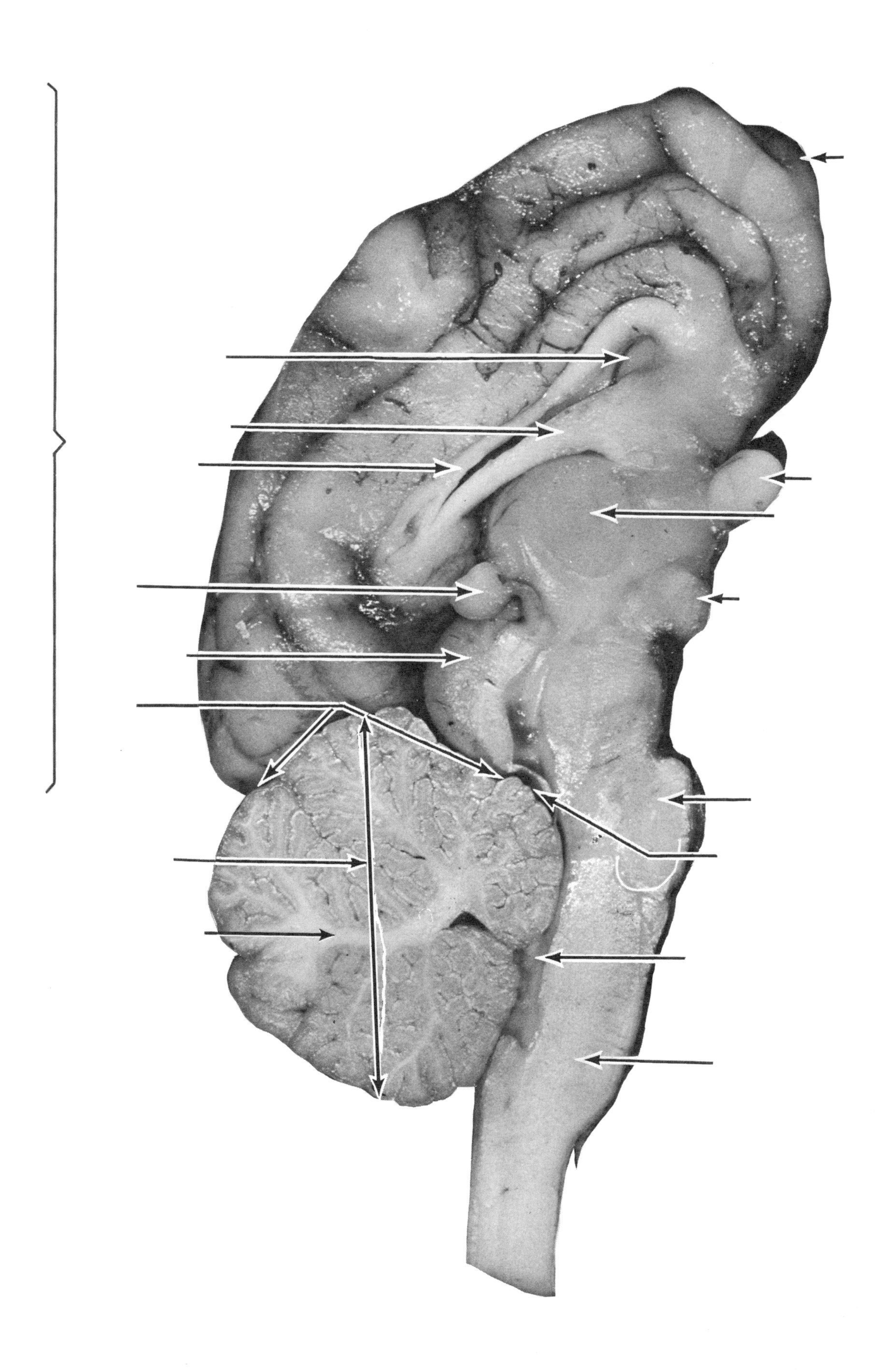

# THE SHEEP EYE - EXTERNAL VIEW AND TRANSVERSE SECTION

The external view of the eye reveals some of the extrinsic *eye muscles* which move the eye. Also seen is the thick layer of *fat* covering the rear of the *eyeball.*

The transverse section reveals the interior of the eyeball. The fat and muscle layers have been removed. The stump of the *optic nerve* may be seen exiting from the rear of the eye.

The *sclera, choroid* and *retina* are seen.

Locate and identify on your specimen all of the structures pictured.

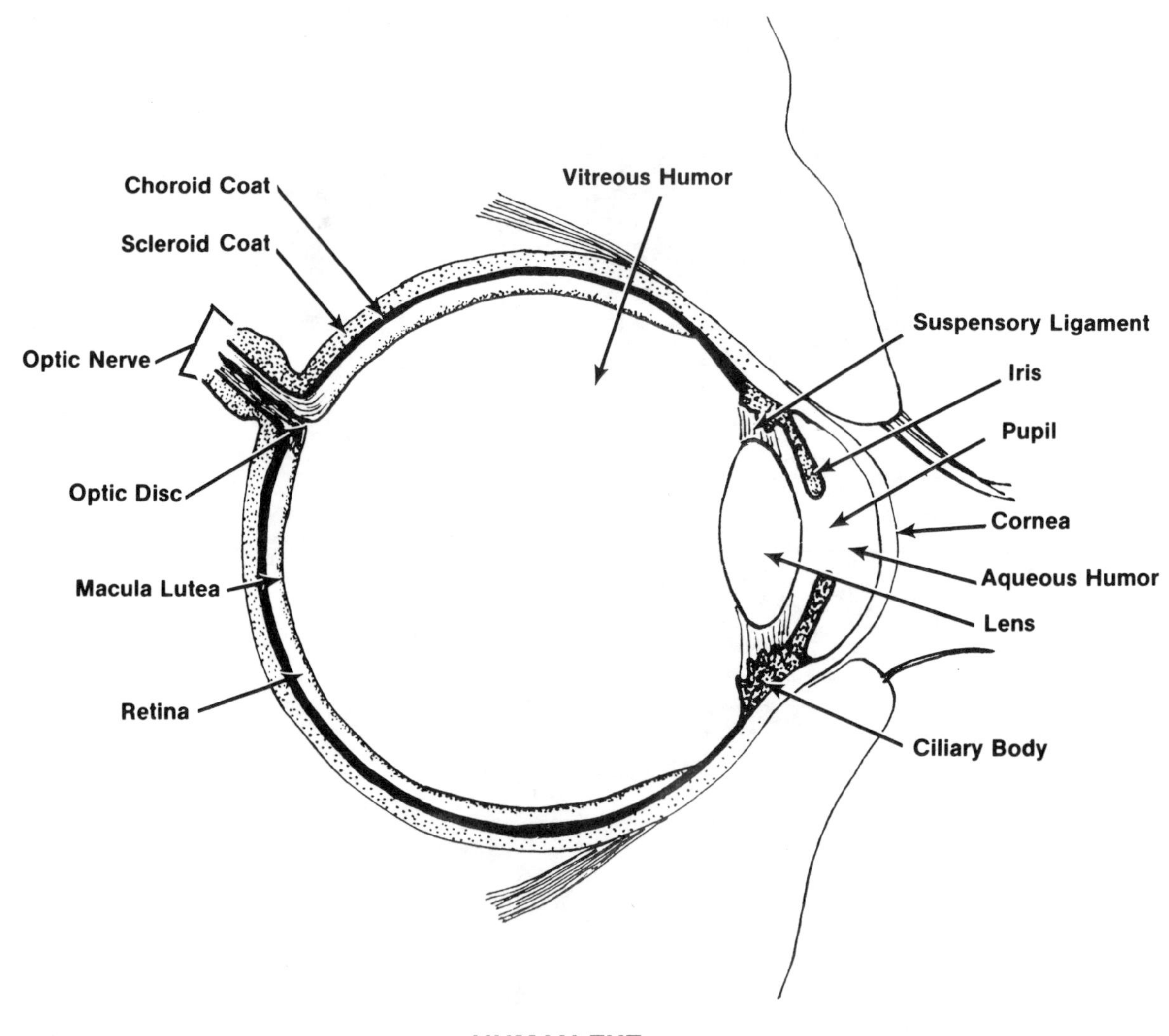


HUMAN EYE

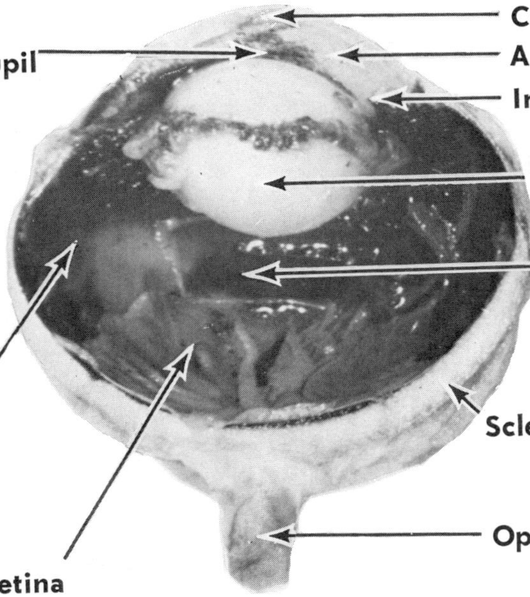

Cornea
Pupil
Anterior Chamber
Iris
Lens
Posterior Chamber
Choroid
Sclera
Optic Nerve
Retina

Transverse Section

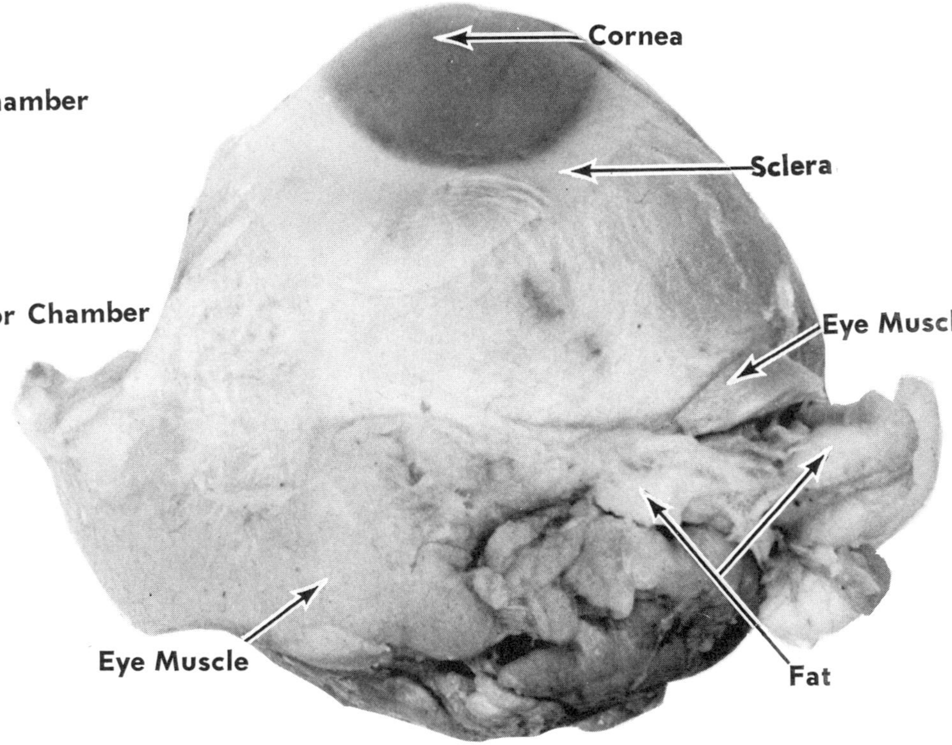

Cornea
Sclera
Eye Muscle
Eye Muscle
Fat

External View

SHEEP EYE

# THE EAR

The ear is extremely difficult to dissect. The structures of the inner ear are very small and are, furthermore, embedded in bone. The dissection can be more readily carried out if that portion of the skull in which the ear lies is first separated from the rest of the head. In addition, the instructor should prepare demonstrations of sagittal sections of the skull.

The following description of the ear is essentially the same for the cat and man. Some of differences are pointed out.

The ear is primarily an organ of hearing and equilibrium. It is divided into three regions, the *external,* the *middle,* and the *internal* ear.

**External Ear**

The external ear consists of the large, externally visible, *pinna (auricle).* This funnel shaped structure is made of supporting elastic cartilage covered by skin. In the cat it is attached by 15 different muscles to the skull. These *anterior* and *posterior auricular muscles* act to move the ear. This enables the cat to detect sounds more readily. By small variations in the movement of the ear the cat is able to avoid enemies and detect its prey.

The *external auditory meatus,* at the proximal end of the pinna, is the opening through which sound waves enter. The *external auditory canal* carries the sound waves toward *tympanic membrane,* or eardrum. The tympanic membrance separates the outer from the middle ear.

**Middle Ear**

The middle ear is the name given to a cavity, the *tympanic cavity,* within a bone, the *tympanic bulla* of the *temporal bone.*

Three small bones are found in the middle ear; the *malleus* (hammer), the *incus* (anvil), and the *stapes* (stirrup). The malleus is attached to the tympanic membrane and articulates with the incus, which, in turn, moves the stapes. The plate of the stapes is attached to a membrane, the *oval window* of the *cochlea.*

The *Eustachian tube* leads from the middle ear to the *nasopharynx.* It serves to equalize pressure against the tympanic membrane. It may also be instrumental in leading an upper respiratory infection towards the middle ear and thereby cause severe earaches.

**Internal Ear**

This is the main area of hearing. Here neurons receive sound waves and sensations of equilibrium and pass them to the brain for interpretation. The structures of the inner ear lie within a *bony labyrinth* within the petrous portion of the *temporal* bone.

The primary receptor of sound, the *organ of Corti* is located within the snail shaped *cochlea.* Fluids within the cochlea are disturbed by sound waves, and these move the delicate sensory hairs. These impulses are passed to the brain by the *auditory nerve,* the eighth cranial nerve.

Sensations of equilibrium and balance are transmitted by nerves originating in three pairs of fluid filled *semicircular canals.* These are oriented in the three planes of space; vertically, horizontally, and obliquely. In addition, two membranous sacs, the *utricle* and the *saccule* are sensitive to changes in equilibrium and balance.

## THE HUMAN EAR

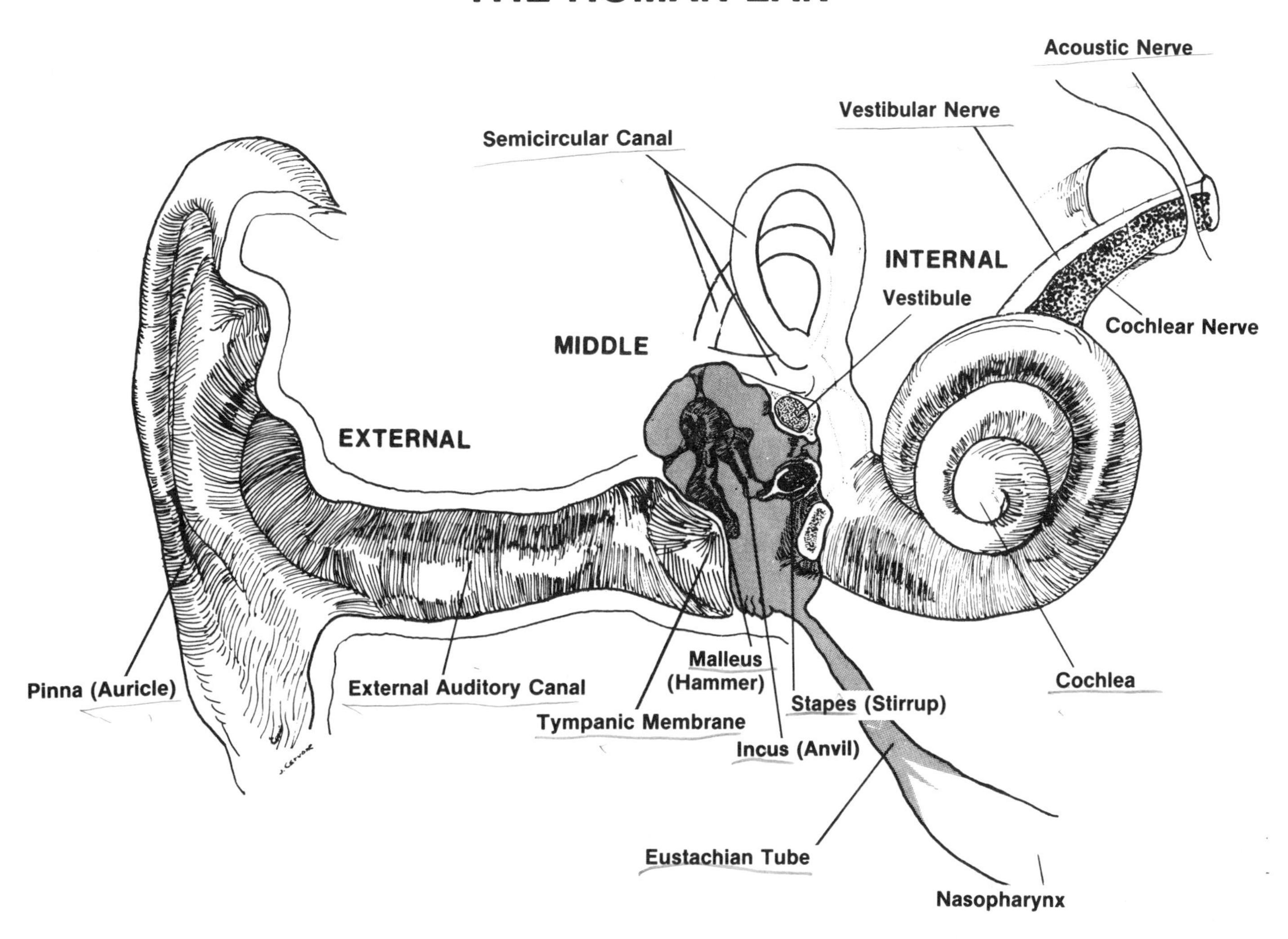


oval window
is below
stapes